Tshilanda Dinangayi Dorothée
Babady-Bila Philippe
Mpiana Tshimankinda Pius

Phytochemical study and anti-sickle cell activity of the genus Ocimum

T shilanda Dinangayi Dorothée
Babady-Bila Philippe
Mpiana T shimankinda Pius

Phytochemical study and anti-sickle cell activity of the genus Ocimum

Imprint
Any brand names and product names mentioned in this book are subject to trademark, brand or patent protection and are trademarks or registered trademarks of their respective holders. The use of brand names, product names, common names, trade names, product descriptions etc. even without a particular marking in this work is in no way to be construed to mean that such names may be regarded as unrestricted in respect of trademark and brand protection legislation and could thus be used by anyone.

Cover image: www.ingimage.com

This book is a translation from the original published under ISBN 978-620-2-28562-9.

Publisher:
Sciencia Scripts
is a trademark of
International Book Market Service Ltd., member of OmniScriptum Publishing Group
17 Meldrum Street, Beau Bassin 71504, Mauritius
Printed at: see last page
ISBN: 978-620-2-57486-0

Zugl. / Approved by: Defended in 2016 at the University of Kinshasa, Kinshasa, Democratic Republic of Congo. Research area: Chemistry of Natural Substances and Medicinal Chemistry

i

"To the King of the ages, Immortal, Invisible, only God, be honor and glory forever and ever! Amen! "1 Timothy 1:17.

To Jesus Christ my Saviour, my Lord and Source of my life;

To my father Crispin Mbuyamba Ndumbi,

To my late mother Gertrude Njiba Tshinanduku,

To my husband Simon Pierre Kanda Mutabazi,

To my children: Carl Kanda Kasongo, Golda Kanda Biseka, Mylène Kanda Njiba, Josaphat Kanda Luangomba, Levi Kanda Ngoyi, Obed Kanda Mbuyamba, Judith Kanda Mbuyi, Divin Benjamin Kanda Bua-Nzambi, Francis Nkuadi Tubajika, Nathan Mukeba Kabulu

I dedicate this work.

Dorothée TSHILANDA DINANGAYI

THANK-YOUS

This work done as part of a doctoral thesis in science would not have been possible without the assistance and support of several people whom I would like to thank in particular.

First of all, I should like to thank Professor Damase Onyamboko for agreeing to direct this work, for his sound advice, availability and rigour.

I would like to thank Professor Philippe Babady-Bila for initiating me in the field of natural substances and for agreeing to co-direct this thesis, having invested himself totally in its realization despite the distance. His remarks, his scientific rigour were a great contribution to the realization of this thesis;

My thanks also go to Professor Pius Mpiana Tshimankinda for his concern to see me move forward, for his unambiguous guidance and the good atmosphere within our research team.

May Professor Jean Paul Ngbolua, one of the pioneers of our research team, find here the expression of my gratitude for his great contribution to the elaboration of this thesis.

I would like to warmly thank Damien Tshibangu Sha Tshibey, Philippe Tsalu, Ben Gbolo, Aristotle Matondo, Domaine Mwanangombo, Joseph Tshidibi, Samy Ngunde and Nicole Misengabu for their scientific contribution and encouragement.

Professors Oscar Shetonde and Pitchou Ngoy, Drs Eddi Dibwe, Jeff Malongwe and Paulin Mutuale, Head of Works, find in this paragraph the expression of my gratitude for having helped me with the spectroscopic and chromatographic analyses of my samples.

My heartfelt thanks to: Professor Joe Ntumba and Daddy Kashishi for agreeing to take the GC and GC/MS measurements of my samples; Dr. Odette Kabena for helping me with the antibacterial tests, and Professor Blaise Mbala for his concern at one point in this work.

I keep a pious thought for Papa Mputu, whom the Eternal One has recovered, for his assistance to the Laboratory of Chemistry of Natural Substances.

I would like to thank Professor Bea and his collaborators at ISTM Kinshasa for their assistance and welcome in their laboratory, for the use of the UV spectrophotometer.

That Professor Bertin Makolo Musuasua, who made my registration at the University possible, then assistant to the Academic Secretary General, finds in these few lines my gratitude.

To all the Professors of the Faculty of Sciences and those of the Department of Chemistry in particular, for all the efforts made for my training, I say thank you.

I would like to express my gratitude to all our colleagues, Assistants and Supervisors, for their advice and scientific contribution.

I would like to thank in particular my husband Simon Pierre Kanda and all my children for their support and encouragement without which this work would not have been possible.

There is a kind couple, a big brother and a big sister for me, who have encouraged me to persevere in this career, informing themselves day after day about the evolution of this work without getting tired. They are Célestin Yanga Kalala and Justine Masengu Kayemba; may they find here the crowning achievement of their expectations and all my gratitude.

I would like to read the joy of the one who bore me in her womb Gertrude Njiba, and of my adoptive parents Joachim Ngoyi and Marie Antoinette Ntanga, who were waiting for this day but alas, the Eternal One has decided otherwise, may His name be praised!!

For the Christian formation and support during the difficult moments of my life, I say thank you to Julienne Malobo Sekele and all the "ministry of reconciliation by the women of fire".

To the Church of God that is in our house, I say thank you for the love and joy that reigns around us and for the spiritual support.

To all my godchildren who wanted to see me reach the end; to Daddy Kanika, Blaise Kanyamanda and all the friends I cannot name here, who gave me your support in various circumstances, I thank you.

I would like to thank all my family and my in-laws for their love and care.

Finally, my gratitude to Professor José Lami Nzunzu and all the members of the jury for agreeing to examine this thesis work.

TABLE OF CONTENTS

ABBREVIATIONS

AA	Ascorbic acid
AcOEt	Ethyl acetate
AR	Rosmarinic Acid
ATCC	American Type Culture Collection
bs	Broad singlet
CJC	Thin Layer Chromatography
CDCl3	Deuterated chloroform
CHCl3	Chloroform
cm	Centimeter
WMC	Minimal Concentration Bactericide
CMI	Minimal Inhibitory Concentration
GSC	Minimum Normalizing Concentration
NCPC	Neuro-Psycho-Pathological Centre
COSY	COrrelated SpectroscopY
GICS/GICS	Gas chromatography coupled with mass spectrometry
CVO	Vaso-occlusive crisis
d	Doublet
DCM	Dichloromethane
dd	Split Doublet
DEPT	Distortionless Enhancement by Polarization Transfer
DMAPP	Dimethylallyl diphosphate
DPPH	2,2-Diphenyl-1-picrylhydrazyl
EAG	Gallic acid equivalent
ED	Dichloromethane extract
EH	Extract with *n-hexane*
EM	Methanol extract
EM/AcOEt(EMA)	Methanol extract in ethyl acetate after washing with water
FDA	Food and Drug Administration
SPM	Geranyl pyrophosphate
H_2SO_4	Sulphuric acid
HbS	hemoglobin S
HCl	Hydrochloric acid
HE	Essential oil
HMBC	Heteronuclear Multiple-Bond Connectivity
HMQC	Heteronuclear Multiple Quantum Coherence
HPLC	High Performance Liquid Chromatography
Hz	Hertz

IC50	50% Inhibitory Concentration
IE	Electronic impact
INERA	National Institute of Agronomic Studies and Research
IPP	Isopentyl diphosphate
IR	Retention Index
J	Coupling constant
LASCHIMED	Laboratory of Natural Substances and Medicinal Chemistry
m	Multiplet
m/z	loaded mass
MeOH	Methanol
N	Normal
Na2S2O4	Sodium metabisulphite
NaCl	Sodium chloride
NaOH	Caustic soda
NH4OH	Ammonia
WHO	World Health Organization
ppm	ppm
RDA	Retro Diels-Alder
DRC	Democratic Republic of Congo
NMR	Nuclear Magnetic Resonance
NMR-13C	Nuclear Magnetic Resonance of Carbon-13
NMR-1H	Proton Nuclear Magnetic Resonance
NMR-2D	Two Dimensional Nuclear Magnetic Resonance
s	Singulet
SM	Mass Spectrometer
t	Triplet
TMS	Teramtehylsilane
TR	Retention time
TSA	Trypticase Soy Agar
TSB	Trypticase Soy Broth
UFC	Unit Format Colonies
UV	Ultraviolet
HIV/AIDS	Human Immunodeficiency Virus/Acquired Immunodeficiency Syndrome
μl	Microliter

SUMMARY

The objective of this study is to determine the chemical composition and evaluate the biological activities (antifalcemic, antibacterial and antioxidant activities) of extracts, essential oils and isolated products of *Ocimum basilicum, O. canum* and *O. gratissimum.*

Chemical screening showed that the three plants had almost the same secondary metabolite profile but at different levels and that saponins were absent in *O. canum.*

Thin layer chromatography carried out on the methanolic extract, followed by HPLC, made it possible to identify in these three plants the flavonoids of kæmpferol type and rosmarinic acid which is the major compound in *O. basilicum,* followed by *O. gratissimum* and rather weak in *O. canum* .

Chromatographic (TLC and column) and spectroscopic (1D-NMR, 2D-NMR, MS) analyses have made it possible to isolate and elucidate the structures of ursolic acid, oleanolic acid and squalene in *O. gratissimum.* In addition, GC/MS identified approximately 35 compounds contained in the same plant, including cyclohexatetracarontane, 1,8-dimethoxy-3-methyl-3-anthraquinone and eicosanol.

35 compounds were identified by GPC and GPC/MS in the essential oil of *O. basilcum,* the main ones being methylchavicol or estragol (35.72%) and linalol (21.25%). 45 compounds were detected in the essential oil of *O. gratissimum of* which five were not identified, the majority compounds being thymol (39.86%), γ-terpinene (22.34%) and *p-cymene (*17.60%). A majority compound has been identified in the essential oil of *O. canum*: it is 1,8-cineole.

To our knowledge, it is the first time that so many chemical compounds have been determined from these plants growing in the DRC.

A few non-aromatic esters with antifalcemic activity have been synthesized. These include esters of palmitic acid, stearic acid and oleic acid.

Pharmacological screening showed that all the plants tested have anti-sickle cell properties, which are due to organic acids (rosmarinic acid, ursolic acid, oleanolic acid). The essential oils extracted from these plants also have an anti-sickle cell activity and thymol is the active principle for the essential oil of *O. gratissimum. In addition,* the non-aromatic esters obtained have shown *in vitro* an antifalcemic activity due to their hydrophobic properties which favour their passage through the cell membrane; this activity has not been observed for oleic acid esters.

To our knowledge, it is for the first time that the antifalcemic activity of these plants, essential oils, rosmarinic acid, ursolic acid, thymol and non-aromatic esters have been studied and observed.

The essential oil of *O. gratissimum* has shown a bactericidal effect on two strains (*Proteus mirabilis* and *Pseudomonas aeruginosa*) and a bacteriostatic effect on *Escherichia coli, Staphylococus aureus, Klesbiella pneumonia and Salmonella typhimurium.*

It also appears from this study that the three plants have an anti-free radical activity, of which that of *O. gratissimum* is superior due to the wide field of action of its extracts and its essential oil.

Key words: *Ocimum*, traditional medicine, medicinal plants, sickle cell anemia, rosmarinic acid, ursolic acid, oleanolic acid, essential oil, thymol.

ABSTRACT

This study aimed to determine the chemical composition and to assess biological activities (antisickling, antibacterial and antioxidant activities) of extracts, essential oils and isolated products from *Ocimum basilicum*, *O. canum* and *O. gratissimum*.

The chemical screening showed that the three plants have almost the same profile in secondary metabolites but in different concentrations and there were no saponins in *O. canum*.

The thin layer chromatography performed on the methanolic extract followed by HPLC allowed to identify flavonoids of kaempferol type and rosmarinic acid which is the major compound in *O. basilicum*, *O. gratissimum* and was less in *O. canum*.

The chromatographical analyses (thin layer chromatography and column chromatography) and spectroscopic (NMR-1D, NMR-2D, MS) allowed to isolate and elucidate the structures of ursolic acid, oleanic acid and squalene from *O. gratissimum*. However, GC/MS allowed to identify around 35 compounds in the same plant of which: cyclohexatetracontane, dimethoxy-1,8 methyl-3 anthraquinone and eicosanol.

In fact, 35 compounds were identified by GC and GC/MS in the essential oil of *O. basilicum* with the main constituents being methylchavicol or estragol (35.72%), linalol (21.25%); and 45 compounds were detected in the essential oil of *O. gratissimum*, five of them were not identified and the main constituents were thymol (39.86%), γ-terpinene (22.34%) and *p*-cymene (17.60%). The main compound in the essential oil of *O. canum* was 1,8-cineol.

To our knowledge, it is the first time that such a huge number of chemical compounds is found in these plants growing in DRC.

Some non-aromatics that have anti-sickling activity were synthetized namely esters of palmitic acid, stearic acid and oleic acid.

The pharmacological profile showed that all the tested plants possess antis-ickling properties. These ones are due to the presence of organic acids (rosmarinic acid, ursolic acid and oleanolic acid). The extracted essential oils of these plants also displayed an anti-sickling activity and thymol is the main active principle of the essential oil from *O. gratissimum*.

before their fifth birthday if they are not medically monitored (Stryer et *al.*, 1997; Voet and Voet, 1998).

In some regions of Africa, carriers of the sickle cell trait represent up to 20% of the population, with a prevalence in Central Africa of 25 to 30% (Voet and Voet, 1998). Two percent of the population in the Democratic Republic of Congo (DRC) are homozygous SS, or more than one million people (Mpiana et *al.*, 2007a).

Sickle cell disease is a major public health problem (WHO, 2006). Several treatment options have been considered to relieve patients. These include bone marrow transplantation, gene therapy, repeated blood transfusions, hydroxyurea, etc. (WHO, 2006). However, these treatments have proven to be not only ineffective and very costly for poor people in Africa, but also blood transfusions predispose patients to a risk of HIV/AIDS infections.

In recent years, several studies have focused on herbal medicines for the management of sickle cell disease (Isamu et *al.* 2002; Moody et *al.* 2003; Ibrahim et *al.* 2007).

Our research team has shown that more than 100 medicinal plants used in traditional Congolese medicine to treat SS anemia, including *Ocimum basilicum, have* anti sickle cell disease activity *in vitro* (Mpiana et al., 2007a; Mpiana *et al.,* 2008, Mpiana et *al.,* 2010).

Our previous results have also shown that this antifalcemic activity is mainly due to anthocyanins (Mpiana et al., 2007b; Mpiana *et al.,* 2008), organic acids such as betulinic and maslinic acid (Tshibangu, 2011); lunaric acid (Ngbolua et *al.,* 2013) and fatty acid esters: butyl stearate (Tshilanda et *al.,* 2014).

0.2. Assumptions

O. basilicum and other species of the genus *Ocimum* have *in vitro* antifalcemic activity according to the chemotaxonomic approach. This activity is due not only to anthocyanins and organic acids but also to other compounds that characterize this botanical genus, notably essential oils.

Essential oils also have antibacterial activity on the strains that sickle cell disease sufferers develop.

The extracts and essential oils from these plants are also endowed with anti-oxidative activity.

The products isolated or identified from these plants, if they are the majority, have antifalcemic and anti oxidative activities.

Synthetic esters also have *in vitro* antifalcemic activity when compared to natural butyl stearate isolated from *Ocimum basilicum.*

0.3. Objectives

0.3.1. Overall objective

The overall objective of our research was to valorize the medicinal plants of the Congolese pharmacopoeia. Cases of *Ocimum basilicum*, *O. canum* and *O. gratissimum*.

0.3.2. Specific objectives

The specific objectives pursued are :
- Chemical screening and screening by thin layer chromatography (TLC).
- Extraction and chromatographic fractionation of the three selected species.
- Structure elucidation of isolated compounds by spectroscopic methods.
- Identification of the chemical composition of extracts, fractions and essential oils by high performance liquid chromatography and gas chromatography coupled to mass spectrometry.
- The synthesis of non-aromatic esters and
- the evaluation of the biological activities of these plants, their essential oils, isolated products and esters.

0.4 Interest

Our work focused on *Ocimum basilicum* L., *O. canum* Sims and *O. gratissimum* L. of the family *Lamiaceae;* and is intended as a contribution to the phytochemical and biological study of medicinal plants belonging to the Congolese flora. It is part of the research program developed by the Laboratory of Natural Substances and Medicinal Chemistry (LASCHIMED) of our Faculty.

0.5. Subdivision

This work is subdivided into three chapters, in addition to the general introduction and the conclusion, as well as the bibliography and annexes.

The first chapter is a review of the literature on sickle cell disease, the genus *Ocimum,* essential oils and some secondary metabolites such as terpenes, fatty acid esters and polyphenols.

The second chapter concerns the experimental part of this study, and presents the materials and methods used.

The third chapter gives the experimental results and their discussion.

CHAPTER I: REVIEW OF THE LITERATURE

1.1. Sickle cell disease

1.1.1. Definition and characteristics

Sickle cell anemia, also called hemoglobinosis S, sicklemia, sickle cell anemia or sickle cell anemia, is a genetic disease, autosomal recessive transmission, linked to a hemoglobin abnormality. Sickle-cell anemia is therefore one of the haemoglobinopathies (Galacteros, 2013).

It was first described by James Herrick, a physician in Chicago, USA in 1910 while examining a 20-year-old black student (Labie & Elion, 2003).

Sickle cell disease is a qualitative abnormality resulting from a mutation leading to the replacement of a polar amino acid (glutamic acid) by valine, an apolar amino acid in the globin beta chains. This pathological haemoglobin is called haemoglobin S (Patrinos *et al.* , 2004). This replacement of a polar amino acid by an apolar amino acid not only changes the oxygen affinity of HbS but also the solubility of this abnormal haemoglobin in the absence of oxygen (Weil, 2005).

The clinical manifestations of sickle cell disease can be summarized in three main categories, namely chronic hemolytic anemia, vaso-occlusive phenomena and susceptibility to microbial infections (Dorn-Beineke & Frietsch, 2002).

In addition to the molecular abnormality that presides over the synthesis of abnormal hemoglobin S in sickle cell disease, other aggravating factors involve, among others, membrane lipid peroxidation of sickle cells and several studies have highlighted the role of free radicals in the physiopathology of this disease (Sess *et al.* , 1992 ; Diatta et *al,* 1999).

Another study has shown that the production of free radicals, due to oxidative stress in patients with sickle cell disease suffering from renal failure, leads to a decrease in the antioxidant activity of serum enzymes (Emokpae et *al.* , 2010).

Thus, complications in sickle cell disease such as chronic hemolysis are the consequence of high oxidative stress and low availability of NO, a vasodilator gas. Hence, in the short term, vasoconstriction which aggravates the consequences of anaemia and endothelial lesions which increase the risk of thrombosis. In the long term, there is chronic arterial damage in the sickle cell patient, with a paradoxical reduction in arterial rigidity and cardiac damage (congestive heart failure) which frequently complicates the patient (Lemgoum et *al.*, 2008; De Viel, 2012).

It should also be noted that attacks of vascular occlusions, which are at the root of many ailments experienced by a sickle cell patient, also cause infections of all kinds that are responsible for a significant portion of mortality and morbidity (pneumopathy, osteomyelitis, meningitis, septicemia). The frequency of infectious accidents decreases with age, but the risk persists throughout life (AFA, 2015).

A 12-year study in Senegal based on a sample of 556 children with sickle cell disease found 67% of cases of vaso-occlusive seizures, 13% of infections and 8% of anaemia. Meningitis, septicaemia and osteomyelitis were the most serious infections in this experiment; *Streptococcus pneumoniae was* the main germ responsible for meningitis, while *Salmonella typhi* and *Staphylococcus aureus* were the only germs isolated in septicaemia and osteomyelitis. However, *Haemophilus influenzae* is also one of the most feared germs in infants and small children (Diagne *et al.* , 2003).

In another experiment, it has been shown that germs are carried in the bloodstream and are often of digestive origin (cholecystitis or gastroenteritis), which explains the frequency of salmonella among the germs responsible (Catonné *et al.* , 2012). The initially aseptic infarction lesion is therefore a privileged area for the implantation and development of salmonella, but also staphylococcus aureus and other bacteria including enterobacteria, *Klebsiella, Escherichia coli,* streptococci and pneumococci (Catonne et *al.* , 2012).

1.1.2. Epidemiology.

Sickle cell disease is found mainly in Africa, but also in Mediterranean countries, the Middle East, India, America (North and South) and Europe. The frequency of occurrence of this disease in Africa varies greatly from one region to another. In central Africa, there is a prevalence of 10-40%, which drops to 1-2% in northern Africa and finally to less than 1% in southern Africa. This variable distribution of sickle cell disease in Africa is explained by the selective advantage that confers the β^S gene with respect to *Plasmodium falciparum, the* causative agent of malaria. It has been observed that *Plasmodium* has much more difficulty in completing its cycle in the weakened red blood cells of a heterozygous sickle cell patient than in a healthy person. However, for homozygous patients, this protection does not exist and malaria is a major cause of mortality in sickle cell disease patients (WHO, 2006).

1.1.3. Physiopathology

Polymerization of hemoglobin S is the basis of the pathophysiology of sickle cell disease. It occurs under conditions of low oxygen pressure and induces the rigidification and deformation of red blood cells into hyperdense sickles as shown in Figure 1.1 below.

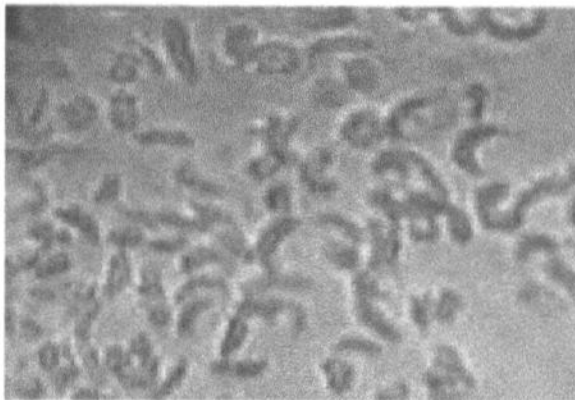

Figure 1.1: Microscopic image of the blood of a homozygous SS patient (Source: Mpiana *et al.*, 2008)

In this image, healthy red blood cells are circular and concave; diseased red blood cells with mutated hemoglobin S, sickle-shaped.

These formed sickle cells play a role in the occlusion of the microcirculation, particularly in the post-capillary venules, leading to the vaso-occlusive crisis (CVO). The deformation and rigidification weaken the red blood cell and can thus lead to its hemolysis. The result is hemolytic anemia (Elion *et al.*, 1999).

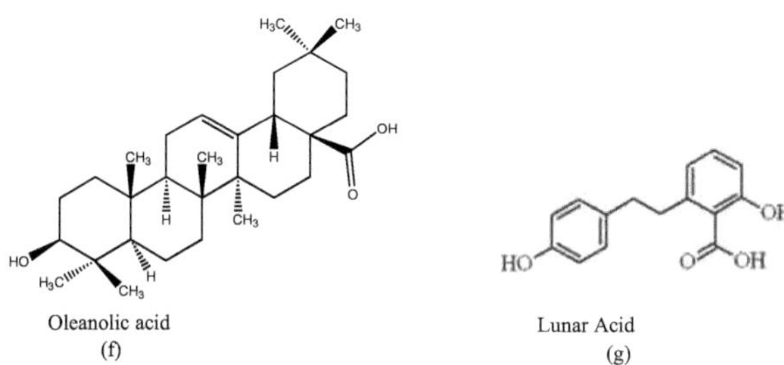

Oleanolic acid
(f)

Lunar Acid
(g)

Figure 1.2: Some anti Sickle Cell Disease molecules isolated from plants.

Finally, it should be noted that some phytomedicines are currently used in human clinics in Africa. This is the case of Nicosan® in Nigeria (Swami *et al.*, 2009), Faca® in Burkina Faso (Nacoulma *et al.*, 2006a), Beat-SS® (Nsimba et al., 2013) and Drépanoalpha® (Ngbolua *et al.*, 2016; Ngbolua et *al.*, 2014) in DRC.

1.2. Plants used

1.2.1. General information on the genus Ocimum

1.2.1.1. Systematic position

The genus **Ocimum** belongs to the family *Lamiaceae* (Lamiaceae), the subfamily *Nepetoideae* (Nepetoidae) and the tribe *Ocimeae* (Ocimeae).

It has about 150 species of herbaceous or bushy plants, annual or perennial, generally aromatic.

As the taxonomy of this genus has undergone frequent upheavals, synonymies are also numerous, and it is not uncommon for the same name to designate, depending on the author, two different species (WCSP Kew, 2012).

According to Lathan & Konda (2007), the two names Ocimum *basilicum and* Ocimum *americanum (Ocimum canum)* are attributed to the same species named Dizulu in Kikongo and Disulu in Tshiluba in the DRC; while a species related to Disulu named Luenyi and collected at Mbuji Mayi in Kasai Oriental has been identified at the Herbarium of the University of Kinshasa as Ocimum *canum Sims.*

1.2.1.2. Phylogenetic classification

The species studied all belong to:

In the Reign	: *Plantae*
At the Junction:	*Spermatophyta*
In clade 1	: *Angiosperma*
In clade 2	: Asteroid
On clade 3	: Lamiidae or Euasterdées I
To the Order	: Lamiales Bromhead (1838)
To the Family	: *Lamiaceae* Martinov (1820)
Au Genre	: *Ocimum*

1.2.1.3. Main species

- ***Ocimum americanum* L.**

  Synonyms: *O. stamineum* Sims, *O. brachiatum* Blume, O. *fluminense* Vell, O. *hispidiculum* Schumach. & Thonn, O. incanescens Mart, O. *pilosum Willd*, *O. citratum Rumph*, *O. hispidum Lam*, *O.* africanum *Lour*, *O. graveolens A.*, *O.* dichotomum *Hochst.* ex Benth. In Cambodia and Thailand, the leaves repel mosquitoes and the mucilage from the seeds is used to thicken soups and desserts (Tropicos.Org., 2016).

- ***Ocimum basilicum* L.** :

  Synonyms: *O. thyrsiflorum L.*, O. *album* L., O. *medium* Mill., O. *bullatum* Lam., O. *integerrimum* Willd., O. *ciliatum* Hornem., O. *barrelieri* Roth., *Plectranthus barrelieri* (Roth) Spreng., O. *caryophyllatum* Roxb., O. *citrodorum* Blanco (Tropicos.Org., 2016).

- ***Ocimum canum:***

  Synonyms: *Ocimum americanum*

  (http://plants.usda.gov/core/profil? Symbol=OCCA4; Berhaut, 1975)

  Noms communs: African mint, basil americain, basil asiatique, hairy basil (http://www.drugs.com/npp/African -mint.html)

- ***Ocimum filamentosum* Forssk**:

  Synonyms: *Becium filamentosum* (Forssk.) Chiov, *Ocimum adscendens* Willd, *Ocimum indicum* Roth, *Plectranthrus indicus (*Roth.) Spreng, Ocimum *cristatum* Roxb, Ocimum *exsul* Collet & Helmsl (http://www.drugs.com/npp/African -mint.html)

- ***Ocimum gratissimum***:

  Synonyms: *O. petiolare* Lam., O. *urticifolium* Roth, O. *viridiflorum* Roth, O. *viridiflorum* Willd., O. *suave* Willd., O. *febrifugum* Lindl., O. *guineense* Schumach. & Thonn., O. cillosum *Weinm., O. paniculum Boj., O. anosurum Fenzl*.

- ***Ocimum hispidus* (Benth) Murata**

13

- *Ocimum kilimandscharicum* **Baker ex Gürke**

- *Ocimum menthaefolium* **Benth**.

- *Ocimum tenuiflorum* **L.** :

  Synonyms: *O. sanctum L.*, O. *monachorum* L., *inodorum* Burm., *O. tomentosum* Lam., *Lumnitzera tenuiflora* (L.) Spreng., *Plectranthus monoachorum* (L.) Spreng, *O. hirsutum* Benth., O. *villosum* Roxb., O. *album* Blanco, *Moschosma tenuiflora* (L.) Heynh., O. *nelsonii* Zipp., O. *virgatum* Blanco.

- *Ocimum ternifolius* (WCSP Kew, 2012)

  In DRC, three species are known: *Ocimum basilicum L, O. canum* Sims and *O. gratissimum* L whose names in some Congolese languages are grouped in Table **1** below:

Table 1. 1: Vernacular names of species studied

Species	Language			
	Kikongo	Lingala	Tshiluba	Swahili
O.basilicum	Dizulu	Musoli	Disulu	-
O.canum	-	-	Luenyi	-
O.gratissimum	Dinsu-nsusu	Lumbalumba	Tshidibuluenyi	Kifukuza-balozi

1.2.1.4. Botanical description

Species of the genus *Ocimum* generally have quadrangular stems, often woody at the base and highly branched. Their leaves are stalked, opposite, membranous, with entire margins or serrate. The terminal inflorescence is simple or branched at the base. Their whorls are clearly interrupted and the cymes are stalkless, unbranched, bearing 3 flowers. The calyx is ovate or campanulate, bilabiate, while the corolla is subegalegal or protruding from the calyx tube. The four stamens are in two pairs of unequal length, declining and protruding (WCSP Kew, 2012).

Figures 1.3. below show the images of three *Ocimums* harvested in the DRC.

Figure 1.3a: *Ocimum gratissimum* (Photo taken in Kinshasa by Tshilanda, 2014)

Figure 1.3.b*: Ocimum basilicum* (photos taken in Kinshasa by Tshilanda, 2014)

Figure 1.3.c*: Ocimum canum* (Photos taken at Mbuji-Mayi by Alain & Muteba, 2014,)

1.2.1.5. Uses

The genus *Ocimum* is often exploited because of its essential oils used in perfumery and the cosmetics industry, which have several biological properties including anti-inflammatory, antipyretic, anti-ulcer, analgesic, deworming and anti-carcinogenic properties (Yamada et al. , 2014; Saliu *et al.* , 2011; Matasyoh et *al.* , 2006).

 Similarly, the aerial parts (leaves and flowers) of these plants are used as stomach aids, lactagogues, mild narcotics, antispasmodics for the digestive tract, diuretics, antimicrobials, dewormers and to repel mosquitoes by their odour (Anonymous 1, 2013; Anonymous 2, 2013; Yamada et *al.* , 2014). In DRC, *Ocimum basilicum* is used to treat SS anemia (Mpiana *et al.,* 2007a; Mpiana *et al.* , 2010), as a condiment in soups, to treat gastritis and haemorrhoids. It is confused with *Ocimum americanum* (Lathan & Konda Ku Mbuta, 2007).

Ocimum gratissimum is used against diarrhea, respiratory diseases, pneumonia, dysentery, vomiting, hepatitis, asthma, and angina. Other works include anti-trypanosomal and anti-plasmodial activities, oral infections and urinary tract infections. Infusions are used because of their tonic, pectoral, headache, flu, cough, dizziness, dermatoses, ear infections and eye infections properties (Anonymous2, 2013; Kpadonou et *al.* , 2014). In Thailand, traditional medicine uses *Ocimum gratissimum* for its antibacterial activity and recently the anti-tumour effect of Ocimum *gratissimum* has been reported on cancer cells while the root decoction of *Ocimum ternifolius* is used against jaundice (Chen *et al.* , 2011).

In India, *Ocimum canum* is used for its medicinal and culinary values; and it is widely used to treat several types of diseases including type 2 diabetes. It is also used to treat fevers, parasitic infections, inflammation, and headaches. Traditional Indian medicine (Ayuverda) recognizes that *Ocimum canum* treats fevers, dysentery, tooth decay and has insect repellent properties. Thus, the herb is known to have antibacterial, antifungal, antiviral and analgesic properties (Anonymous 3, 2013).

1.2.1.6. Some previous work on the genus Ocimun

(a) *Ocimum basilicum*

Numerous articles have been published on different species of the genus *Ocimum* for their multiple uses as condiments, as insect repellents, in perfumery and traditional medicine, and as sources of essential oils (Yusuf et *al.*, 1994).

For *Ocimum basilicum,* the oils are a mixture of linalool and methylcharvicol and/or 1,8-cineole as a base composition. The other most common compounds are: eugenol, camphor, thymol, β-caryophyllene and citral (James *et al.* , 1999).

Dry leaves of this kind contain 0.20-1% essential oil. The composition varies considerably according to the medium (Takano, 1993; Van Duong, Nguyen, 1993).

Six triterpenic acids were identified in the dichloromethane extract of the roots of genetically modified *Ocimum basilicum from* Egypt. They are betulinic, oleanolic, ursolic, 3-epimaslinic, alphitolic and euscaphic acids and have shown a hepatoprotective effect (Marzouk, 2009).

The phenol composition and polyphenoloxidase activity (PPO) of *Ocimum basilicum* were determined at two stages of maturity and after different drying conditions. Two families of phenolic compounds were detected, hydroxycinnamic derivatives and flavonols. In the first family, rosmarinic acid is the most abundant. Among the flavonols, compounds such as rutin, kaempferol-3-rutinoside, quercitrin and isoquercitrin have been identified (Baritaux *et al*, 1991).

b) *Ocimum gratissimum*

For *Ocimum gratissimum* oil, several studies show that thymol (43.45%), para-cymene (12.26%) and γ-terpinene (12.11%) are majority compounds (Yayi et *al.* , 2001). Another study conducted in Kenya on the essential oil of the leaves of this plant indicates that in this species, the majority compounds are eugenol (68.8%), methyl eugenol (13.21%) and *cis-ocimene (7.47%)* (Kpodekon, et *al.* , 2014).

Further research on the essential oils of this plant has revealed different proportions of chemical compounds around the world. In Cameroon, one study indicates thymol (47.7%) and β-terpinene (14.3%) as the majority compounds (Kpodekon, *et al*, 2013).

Some research has also highlighted the predominance of thymol (70.8%) in the essential oil of *Ocimum gratissimum* from the Ivory Coast (Sahouo *et al.* 2003).

In the same country, another study has obtained in the essential oil of *Ocimum gratissimum*, thymol (34.6%), p-cymene (25.2%), α-selinene (6.8%), myrcene (5.4%), *(E)*-β-karyophyllene (4.9%) and α-thujene (4.5%) (Oussou et *al.* , 2004).

A team in Nigeria found eugenol (61.9%) and *cis-cimene (*8.2%) in the same plant (Saliu et *al.,* 2011).

An extract of *Ocimum gratissimum* showed an *in vivo* effect against *Phytophtlora palmivora, the* causal agent of *Cocoa* black spot disease (Awuah, 1994).

The aqueous extract of *O. gratissimum* showed an anticancer effect and an effect against the neoformation of blood vessels (cancers) in a study in the USA (Nangia-Makker et *al.* , 2007).

C) *Ocimum canum*

The essential oils of *O. canum* from two regions of the island of Grande Comore showed some differences in their chemical composition and showed antibacterial and antifungal activity: the one from the region of Maweni Dimani has a higher yield than the one from the region of Ivoini Mitsahouli. The former contains 1-8 cineole (48.88%), camphor (14.98%), α-pinene (5.71%), β-pinene (4.66%) and γ-eleomene (3.91%) as major constituents while the latter contains the other, *The* majority constituents are 1-8 cineole (34.22%), camphor (13.69%), isopropyl propanoate (9.13%), γ-elemene (5.43%) and α-pinene (3.83%) (Hassani *et al.* ,2011).

In Ghana, Nyarko *et al.* (2002) showed that *O. canum is* endowed with antidiabetic properties *in vivo.*

The study of essential oils of *Ocimum basilicum, Ocimum canum, Ocimum gratissimum* harvested on 14 different sites in Togo was carried out. The essential oil of *O. basilicum* contains mainly estragole or the linalol-estragole mixture. In the essential oil extracted from *Ocimum canum*, terpin-4-ol, linalol, γ-terpinene and sesquiterpene hydrocarbons (β-caryophyllene, *(E)*-α-bergamotene and bicyclogermacrene) are majority compounds while thymol, γ-terpinene and *p-cymene* are identified as abundant compounds in the essential oil of *Ocimum gratissimum* (Sanda *et al.* , 1998).

Several studies have shown the presence of triterpenoids and in particular ursolic, oleanolic, betulinic acids etc. which have been isolated from several species of the genus *Ocimum* including *Ocimum basilicum, O. gratissimum* (Goretti et al. , 2008).

d) *Ocimum sanctum*

A few compounds were isolated from an extract of fresh leaves and stems of *Ocimum sanctum* that showed antioxidant activity. They are cirsimaritin, isothymusin, isothymonin, apigenin and rosmarinic acid (Kelm *et al.*, 2003).

A positive correlation was demonstrated during a study carried out in Iran between antioxidant activity and the amount of phenolic acids contained in some species of the genus *Ocimum* (Javanmardi et *al*, 2003).

In India, betulinic acid has been isolated from cultures of different species of *Ocimum* for its anticancer effect (Pandey *et al.* , 2015).

1.3 Essential oils and some secondary metabolites

1.3.1. Essential oils

1.3.1.1. Definitions and composition

Essential oil (HE), or sometimes vegetable essence, is the hydrophobic liquid of a plant's volatile odoriferous compounds. It is obtained by steam distillation (or hydro-distillation) of various plant materials such as flowers, leaves, stems, roots or whole plants. Unlike essential amino acids, "essential" oil does not mean "essential" in the sense of: indispensable, but rather means "the quintessence of" (Roudmitska, 1980; Bruneton, 1987; AFSS, 2008).

Essential oils consist mainly of two groups of odorous compounds that are distinct according to the metabolic pathway used. These are terpenes, which are preponderant in most essential oils, and phenylpropane derivatives, which are found as the majority compounds in some of these oils. We also find in these essences various other minority constituents, and many derivatives with various functions that are also considered terpene compounds.

The word "terpene" comes from **"turpentine"** (from the Latin *balsamum terebinthinae*). Turpentine contains "acid resins" and some hydrocarbons. Terpenes are hydrocarbons, but this group also includes oxygenated derivatives such as alcohols, aldehydes, acids or ketones, hence the general name *terpenoids* (Ruzicka & Wallach, 1887).

Their structures are formed according to the general principle according to units generally called **isoprene** units, $CH_2=C(CH_3)-CH=CH_2$, or (C_5H_8), forming the carbon skeleton of terpenes. This is the rule for isoprene found by Ruzicka & Wallach (1887).

Isoprene (methylbuta-1,3-diene) is also known as hemititerpene.

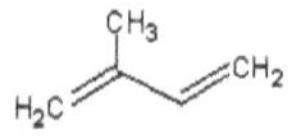

Figure 1.4: Hemiterpene.

Consequently, terpene hydrocarbons have molecular formulae of type $(C_5H_8)_n$, where *n is the* number of isoprene units classified in Table 1.2 below.

Table 1.2: Classification of Terpenes

Terpenes	Number of isoprene units
Hémiterpène	1
Monoterpène	2
Sesquiterpene	3

Diterpene	4
Triterpène	6
Tetraterpene	8
Polyterpène	more than 8

Monoterpenes (n=2): these terpenes are C10 hydrocarbons (C10H16). They can be acyclic (myrcene, ocimene), monocyclic (α and γ-terpinenes) or bicyclic (pinene, camphene).

The structures of a few monoterpenes are grouped together in the following figure 1.5.

- Acyclical:

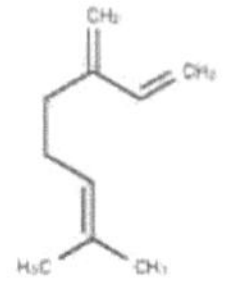

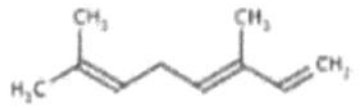

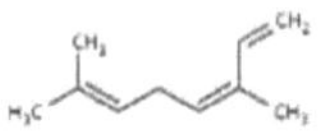

Myrcene   (E)-β-Ocimene   (Z)-β-Ocimene

- **Monocyclic:**

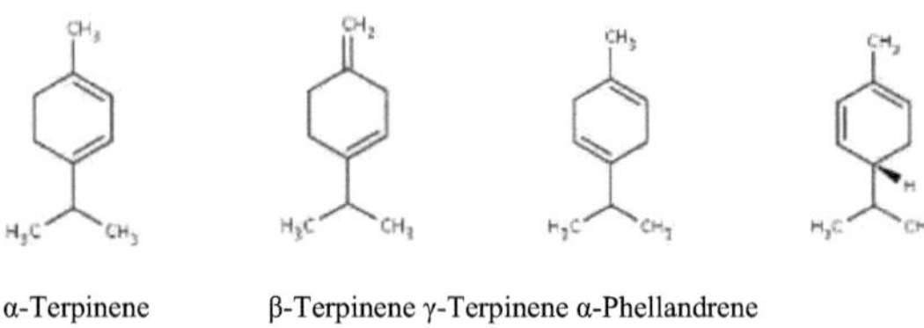

α-Terpinene   β-Terpinene γ-Terpinene α-Phellandrene

- **Bicyclicals:**

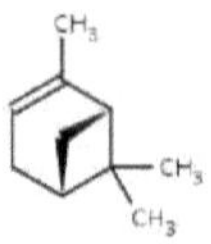

α-Pinene 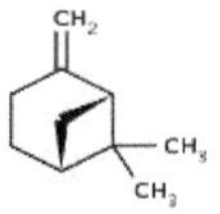   β-Pinene 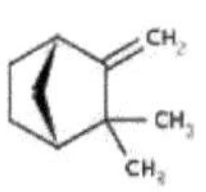   Camphene 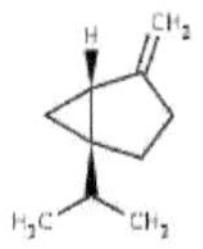   Sabinene

Figure 1.5: Molecular structures of monoterpenes. Source: (ChEBI, **2015)**.

To this group of terpene compounds are added derivatives with special chemical functions such as alcohol, aldehyde or ether (Fig.1.6).

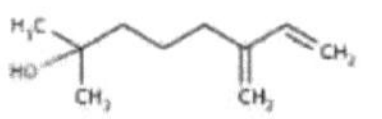

Myrcenol 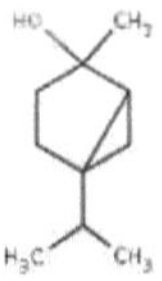   Sabinene Hydrate

Figure 1.6: terpene derivatives with special chemical functions

- Sesquiterpenes (n=3): these are C15 hydrocarbons, i.e. one and a half times (= sesqui) the molecule of terpenes (C10H16). They can be linear (farnesene) or cyclic (β-bisabolene) (Fig.1.7.) (Breitmaier, 2006).

o **Linear** :

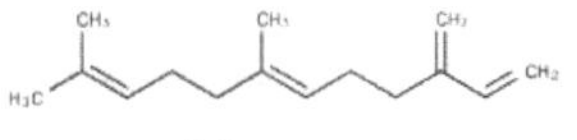

trans-β-farnesene

o **Cycliques**

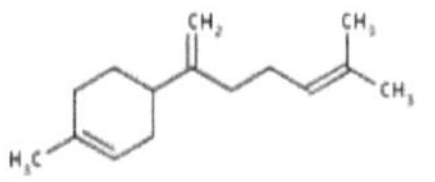

β-bisabolene

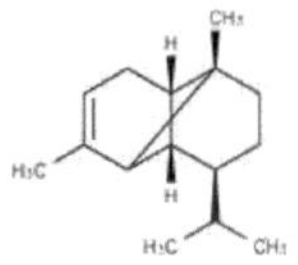

α-copaene

Figure 1.7: Sesquiterpenes

The terpenoids are all formed by the condensation of a phosphorylated hemititerpene (isopentyl diphosphate: IPP) and dimethylallyl diphosphate (DMAPP) to give geranyl pyrophosphate (GPP) (Fig. 1.8.) (Breitmaier, 2006).

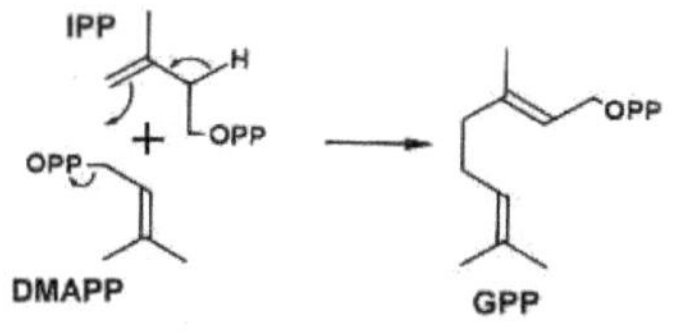

Figure 1.8: Formation of geranyl pyrophosphate

For higher plants, PPI formation follows the biochemical pathway of mevalonic acid (Breitmaier, 2006).

Mono- and sesquiterpenes are the main constituents of essential oils while other terpenes are the constituents of resins, balsamines, waxes and rubber (Breitmaier, 2006; Rhourri-Frih, 2009, Cyberlipid Center, 2015).

1.3.1.2. Natural state

Essential oils are almost exclusive in the branch of spermatophytes which are capable of elaborating these volatile principles and are in first approximation grouped in a rather limited number of families of Angiosperms belonging to the orders of Magnoliales, Lorales, Rutales, Lamiales, Asterales.

Quantitatively, the essential oil contents are very low, often less than 1%; high contents such as those of the bud, clove (15%), are exceptional (Anonymous 4, 2012).

Essential oils are liquid at ordinary temperatures and are only very rarely coloured. Their density is generally lower than that of water. They are almost always endowed with rotational power and have a high refractive index. They are soluble in alcohols and common organic solvents but not in water; they are fat-soluble and can be entrained in water vapour. They are volatile (Babady, 1986; Bruneton, 1999).

1.3.1.4. Medicinal properties

Many essential oils have medicinal properties that have been used in traditional medicine since ancient times and are widely used today. They have anti-infectious, functional rebalancing, psychic rebalancing and endocrine rebalancing properties, depending on their specific components (Roudmitska, 1980; Babady, 1986; Bruneton, 1987; Anonymous 4, 2012).

Thus the properties offered by essential oils are numerous; analgesic (chamomile, pine, lavender, mint), tonic (mint, black pepper, rosemary, sandalwood, sage), calming (chamomile, myrrh, orange), anti-inflammatory (lemon, frankincense, geranium), regulating (chamomile, lavender), healing (eucalyptus, lavender, hyssop, rosemary), and much more (Shigegarul et *al.* , 2006).

1.3.1.5. Chemical composition of essential oils of Ocimum

The chemical composition of the essential oils extracted from each species of *Ocimum* is variable. Different chemotypes (or chemotypes) have been described in the literature. Nine chemotypes have been reported for *Ocimum gratissimum*.

These are the following chemotypes:

- Linalol/methyl chavicol (Prota4u, Nov.2015).
- Eugenol/1,8-cineole/sesquiterpenes (Prota4u, Nov.2015; Lawrence, 1992).
- Methyl cinnamate (Lawrence, 1992).
- Methyl-eugenol/eugenol (Fun and Baerheim, 1990).
- Ethyl cinnamate (Vostrowsky et *al.*, 1990).
- Citral (Ali and Shamsuzzaman, 1968).
- Geraniol (Hegnauer, 1966; Charles and Simon, 1992).
- Eugenol-rich chemotype (Yusuf *et al.*, 1998; Lawrence *et al.*, 1980).
- Chemotype rich in thymol (Yusuf *et al.*, 1998; Lawrence *et al.*, 1980).

The chemotype rich in eugenol and the one rich in thymol are the most commonly encountered.

The literature also reports the presence of several chemotypes of *Ocimum americanum* Linn (synonym: *Ocimum canum* Sims) based on the chemical composition of its essential oil.

These are the following chemotypes:

- Fenchone   (Yusuf *et al.*, 1998)
- Citral (Lawrence *et al.* , 1980)
- Eugenol (Choudhary *et al.*, 1989)
- Terpineol (Ekundayo *et al.*, 1989)

- 1,8-cineole (Chalchat *et al.* ,1996)
- β-caryophyllene /(E)-α-bergamotene/bicyclogermacrene (Zollo *et al.* , 1998)
- Methyl cinnamate (Sanda *et al.,* 1998)
- Methylchavicol/α-terpineol (Martins *et al.,* 1999)
- Camphor (Chalchat *et al.* , 1999)
- Linalol (Chagonda *et al.* , 2000)
- Limonene (Yayi *et al.,* 2001)

It is also reported in the literature 7 chemotypes for *Ocimum basilicum*, namely :

- Linalol
- Linalol-eugenol
- Methyl chavicol
- Methylchavicol-linalol
- Methyl eugenol-linalol
- Methylcinnamate-linalol and
- Bergamotene (Valtcho *et al.,* 2008).

1.3.1.6. Uses of terpenes

The uses of terpenes are very numerous: they are used as perfumes, medicines, dyes, etc.

Most terpenes are the odoriferous principle of plants; this is the case of the ocimene of basil, the myrcene of laurel, the limonene of lemon, the pinene and pine .

Many terpene molecules have antiseptic properties (clove, thymol, eucalyptol, camphor, etc.). Historically, terpenes have been widely used for embalming, hence the term *"balsamic"* given to plants and oils containing them (Anonymous 5, 2015).

❖ **Some terpenes and their uses**

Eucalyptol is a monoterpenoid abundant in nature, it is found in eucalyptus, tea tree, bay leaves, basil and sage; it is well known for its antiseptic, antibacterial and anti-inflammatory properties.

γ-terpinene, a monoterpene found in various citrus fruits and herbs such as oregano and marjoram, is reputed to have antioxidant properties.

Phytol, a diterpenoid used by insects as a deterrent against predators, is also used in various household products such as detergents and soaps.

β-caryophyllene: it is a sesquiterpene found in cloves, rosemary and hops, it produces anti-inflammatory effects and has been shown to act as a selective agonist of the CB_2 receptor (no other terpene or terpenoid has been shown to influence CB receptors).

Karyophyllene oxide is the substance identifiable in cannabis by drug-sniffing dogs.

Nerolidol: present in neroli, ginger and jasmine, is a sesquiterpenoid with a fresh, woody fragrance; it is currently being studied both as a facilitator for transdermal drug delivery (due to its ability to penetrate the skin) and as an inhibitor of the protozoan Leishmania.

Guaiol is a sesquiterpenoid found in cypress and *Guaiacum* (a genus comprising five slow-growing trees and shrubs native to tropical America); *Guaiacum* itself has been used for centuries to treat coughs, arthritis and syphilis.

Eudesmol, another sesquiterpenoid, is used as a fixative in perfumery, while trans-β-farnesene acts as a natural insecticide in many plant species, including potatoes (Seshata, 2014).

Guanacastapene is a diterpene with interesting antibiotic properties and taxol, extracted from yew, has remarkable anti-cancerous properties (Davoust, 2015).

Squalene is considered a dietary supplement usually found in shark liver in the form of oil (such as olive oil). Taken orally, squalene is considered to be completely safe compared to squalene taken by injection, which is the subject of controversy. High quality, natural squalene, free of trace metals including mercury and other toxins, is considered a powerful antioxidant and is believed to be beneficial to health (Kelly, 1999; Das & Baruchel, 2000; Mercola, 2009;). This molecule is used in cosmetology in moisturizing creams as an agent that penetrates the skin quickly without leaving any oily traces or sensations on the skin and blends well with other oils and vitamins. Squalane, which is a saturated form of squalene being less sensitive to oxidation, is more commonly used in cosmetics than squalene. Toxicological studies have shown that at the concentrations used in cosmetics, both squalene and squalane have low toxicity and are not irritating or sensitizing to human skin (Anonymous 6, 1982).

Squalene is used in the medical field as a vaccine adjuvant in the form of an emulsion added to the vaccine substance to make the vaccine more immunogenicand has been used since 1997 in an influenza vaccine at a rate of approximately 10 mg of squalene per dose.

This vaccine has been approved by the health agencies of several European countries, but not by the Food and Drug Administration (FDA), which has not authorized its use in the United States (WHO, 2015). [

Squalene is one of the causes of overfishing of sharks, which are hunted for their fins, but whose livers are beginning to be sold to produce capsules supposedly good for health.

Environmental (strong decline of sharks) and health concerns (fish liver also stores toxins of health concern) have motivated its extraction from plants (EWG, 2015).

1.3.1.7 Some previous work on terpenes

Several studies are being conducted on this class of natural compounds. This is the case of betulinic acid, a triterpene of the lupeol type that has shown *in vitro* anti sickle cell activity (Tshibangu, 2011).

Several studies have proven that:

- Ursolic acid has a large number of biological activities, namely: anti-inflammatory, hepatoprotective, analgesic, antimicrobial, antimycotic, virostatic, antidiabetic, anti-malarial, antioxidant. It is also an immunomodulator, diuretic, antispasmodic, anti-athrosclerotic, anti-tumour, anti-HIV, inhibits the growth of *Mycobacterium tuberculosis and* has anti-leishmaniac activity (Moulisha et al., 2010; Babalola & Shode, 2013; Wang et *al.,* 2014).

- Oleanolic acid also has a large number of biological activities such as its isomer (ursolic acid): It has tonic and antioxidant effects, protects the cardiovascular system, has a vasodilating effect, by direct action on the vascular endothelium, promotes the regeneration of endothelial tissue, influences the release of nitric oxide (modulator of vascular tone) and prevents various diseases such as hypertension, atherosclerosis, myocardial infarction and heart failure. It has anticancer and antitumor, radio and chemo-protective, anti-inflammatory, antimicrobial, immunomodulating, antiviral and anti-C3-convertase effects (Kapil et *al,* 1994; Jiménez-Arellanes *et al,* 2013;).

1.3.2. Polyphenols

1.3.2.1. Natural state

Polyphenols are a family of organic molecules widely present in the plant kingdom. They are characterized, as the name indicates, by the presence of one or more phenolic groups (comprising at least one aromatic nucleus bearing one or more hydroxyl groups with or without other functions). (Sarni-Manchado and Cheynier, 2006). They can range from simple molecules, such as phenolic acids (gallic acid), to highly polymerized compounds of more than 30 kDa, such as tannins (tannic acid). Some examples of structures are given below (Fig.1.9.):

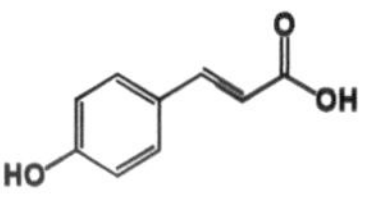

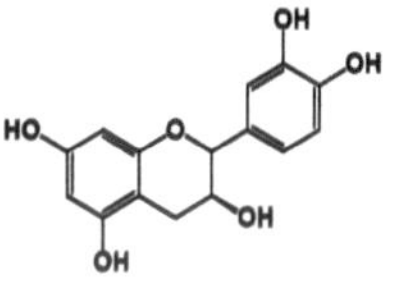

4-Hydroxy-4-cinnamic acid Epicatechin

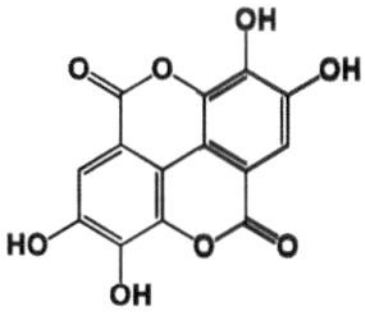

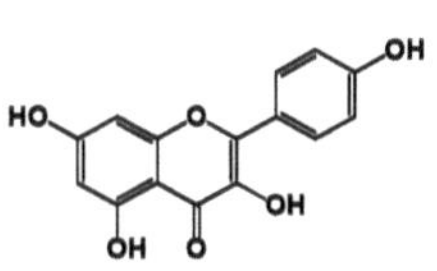

Ellagic acid Kæmpferol

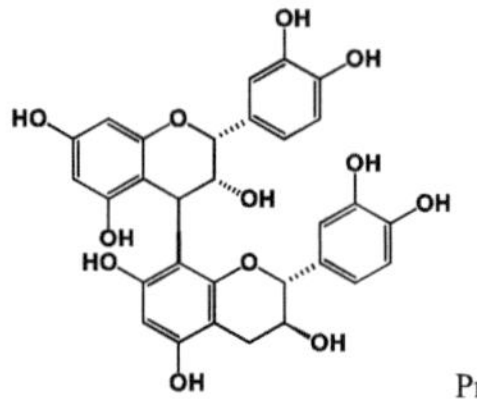

Figure 1.9: Molecular structures of some phenolic compounds

Polyphenols are becoming increasingly important, not least because of their health benefits. Indeed, their role as natural antioxidants is attracting increasing interest in the prevention and treatment of cancer inflammatory cardiovascular [] and neurodegenerative diseases. They are also used as additives for the food, pharmaceutical and cosmetic industries (Orgogozo et *al.* 1997; Manach *et al.*, 2004).

1.3.2.2. Classification

Polyphenols are commonly subdivided into simple phenols, acid-phenols (derived from benzoic or cinnamic acid) and coumarins, naphthoquinones, stilbenoids (two C6 rings linked by 2C), flavonoids, isoflavonoids and neoflavonoids and polymerised forms: lignans, lignins, condensed tannins. These basic carbonaceous skeletons are derived from the secondary metabolism of plants, elaborated by the shikimate pathway (Dewick, 1995).

Based on the basic carbon structure, the following main classes of phenolic compounds can be identified (Table 1.3.):

Table 1.3: Main classes of phenolic compounds, with an example of a class, formula and origin

Carbon skeleton	Class	Formula	Origin
C6-C1	Acids hydroxybenzoic acids	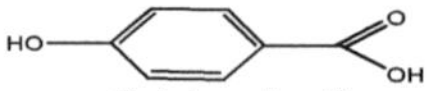 *p-Hydrobenzoic* acid	Spices, strawberries

C6-C3	Hydroxycinnamic acids	*p-Coumaric* acid	Tomatoes, garlic
	Coumarines	Ombelliferone	Carrots, coriander
C6-C4	Naphtoquinones	Juglon	Walnuts
C6-C2-C6	Stilbenoids	*Trans-resveratrol*	Grape
C6-C3-C6	Flavonoids	Kaempferol	Strawberries
	Isoflavonoids	Daidzein	Soybeans
	Neoflavonoids	Dalbergichromene	Dalbergia sissoo
(C6-C3)₂	Lignanes	Enterodiol	Intestinal Bacteria

(Source: Bruneton, 2009; Laughton et *al.*, 1991)

1.3.2.3. Previous work

A study carried out in Côte d'Ivoire on 10 plants including *Ocimum gratissimum* showed that the leaves of *Ocimum gratissimum* are richer in total phenols (7818.66±265.05 µg gallic acid equivalent EAG/g), a result expressed in µg per gram of dry matter, rich in total flavonoids (18.02±4.78%) and flavonic aglycones (0.73±0.17 mg/g) (N'Guessan et *al.* , 2011).

Another study showed that polar extracts (aqueous, methanolic and ethyl acetate) of *Ocimum basilicum* showed *in vivo* anticholesterol and antiglycemic activity and that the polyphenol content is directly proportional to the polarity of the extraction solvent (Harnafi et *al.*, 2008).

1.3.3. Natural fatty acid esters

1.3.3.1. Natural state

Esters are widespread in nature; those derived from monoalcohols and relatively short mono-carboxylic acids are volatile and odorous; they are found in essential oils and in fermented beverages (wine, beer, cider). Others, belonging to the group of fatty substances, very numerous in animal and vegetable fats and oils, are triesters of the trihydric alcohol glycerol and three identical or different fatty acids. These triglycerides are important raw materials in the chemical industry and are an essential part of human and animal nutrition (Mordret & Helme, 1975, Dupuis, 2013).

They are responsible for the flavour and fragrance of many fruits. Food flavourings involve many natural (or identical to natural) esters and synthetic fragrances also contain a large number of them. For example, methyl formate smells like rum; isobutyl acetate smells like bananas; methyl butyrate smells like apples; ethyl butyrate smells like pineapples; and isoamyl butyrate smells like pears. A raspberry flavour formula uses nine esters, two carboxylic acids, acetaldehyde, glycerol and ethanol (Mordret & Helme, 1975; Dupuis, 2013).

Esters are also found in waxes. There are waxes of both plant and animal origin. Beeswax for example contains mainly myricyl palmitate $C_{19}H_{31}CO_2C_{31}H_{63}$) (Mordret & Helme, 1975; Dupuis, 2013).

1.3.3.2. Industrial uses of esters

Synthetic esters are used in perfumery. Some mono- and diesters of middle alcohols are used as plasticizers in the formulation of plastics.

The ester function is present in many synthetic macromolecules, including textile fibres such as polyglycol terephthalate (Tergal), films such as cellulose triacetate or thermosetting resins such as polyglycerol orthophthalate, which are basic products of the paint industry.

The purely chemical interest of esters is no less considerable, they are endowed with a moderate reactivity that allows some of them to be used as reaction or extraction solvents (Morris Kates, 1986).

Hydrolysis of triglyceride ester linkages results in saturated or unsaturated fatty acids, the most important of which contain 12 to 22 carbons (Malumba, 2007).

1.3.3.3. Previous work on fatty acid esters and sickle cell disease

Several studies have shown that ethyl palmitate (in association with snake venom) inhibits the action of cells of the reticuloendothelial system in normal rats injected with whole blood from sickle cell subjects by protecting these diseased erythrocytes against phagocytosis, early hemolysis and falciformation under hypoxic conditions (Morag et *al.*,1970; Oswaldo *et al.*,1973).

One study demonstrated the antifalcemic activity of amino acid benzyl esters (Gorecki et *al.*, 1980).

After this brief review of the literature which gave the idea about sickle cell disease and its treatment methods, the genus *Ocimum* and previous work, essential oils and some secondary metabolics characterizing this genus, the following chapter presents the materials and methods used to carry out this work.

CHAPTER II: MATERIALS AND METHODS

2.1. Biological material

2.1.1. Plant material

The plant material consists of the leaves of *Ocimum basilicum* L, *O. canum* Sims and *O. gratissimum* L.

As these are cultivated plants, they were harvested at several locations in Kinshasa city (Ndjili Brasserie, CNPP Valley and Quartier Mandela) for *Ocimum basilicum and O. gratissimum*. *O. canum* was harvested in Mbuji-Mayi chief town of Kasai-Oriental province.

O. basilicum has been authenticated by the exsiccatum of herbarium no. 425/Devred, located at the Faculty of Sciences of the University of Kinshasa.

O. canum has been identified by herbarium exsiccatum no. 686/A.Thieband while *O. gratissimum* has been identified by herbarium exsiccatum no. 8016/R. Dechamps; all kept at the herbarium of the Institut National des Etudes et Recherches Agronomiques (INERA) located at the Faculty of Sciences of the University of Kinshasa.

The plants were dried in ambient air (about +27°C) at the Laboratoire de Chimie des Substances Naturelles et Chimie Médicinale for about two weeks and then pulverized with a commercial Waring (R) grinder.

The fresh leaves of these plants, were used for the extraction of essential oils with the Rotamag CRV14 heating cap, by hydrodistillation.

2.1.2. Non-plant material

The blood used in this study was collected from patients undergoing preventive treatment at the Centre de Médecine Mixte d'Anémie SS de Yolo in Kinshasa. The test sample contained in a test tube was directly combined with a standard anticoagulant, sodium citrate, and then stored in a refrigerator at ± 4°C.

2.2 Methods

2.2.1. Chemical screening

Chemical screening was carried out on aqueous and organic extracts (Gazengel *et al.* , 1999).

For this study, research was conducted on aqueous macerate of the leaves of three species of *Ocimum* (*O. basilicum, O. canum* and O. *gratissimum)* and the dichloromethane extract.

2.2.1.1. Preparation of the aqueous extract

To 5 g of leaf powder of each plant, add 100 ml of distilled water, leave to macerate for 24 h, then filter; the filtrate constitutes the macerate.

➢ *Tests performed on the aqueous extract*

1. Search for polyphenols

Take 1 ml of the aqueous extract in a test tube, add a few drops of Burton's reagent. In the presence of polyphenols, the solution turns intense blue (sometimes accompanied by a precipitate). Then systematically look for the different phenolic compounds (flavonoids, anthocyanins, leucoanthocyans, tannins, quinones).

2. Search for flavonoids

In a test tube, put about 1 ml of the aqueous extract, add a few drops of the Shinoda reagent with a few magnesium shavings. Then add a few drops of isoamyl alcohol. The formation of a thin film of orange (flavones), red (flavonols) or purple (flavonones) colouring indicates the presence of flavonoids.

3. Search for anthocyanins.

To 3 ml of aqueous extract add 2 ml of 20% HCl and leave to stand for 10 min. In the presence of anthocyanins, a purplish-red coloration of anthocyanin chloride develops which may crystallize.

4. Testing for leuco anthocyanins.

To 1 ml of aqueous extract, add a few drops of Shinoda reagent and isoamyl alcohol and heat in a water bath. A purplish-red coloration in the supernatant layer is a positive test.

5. Search for tannins

1 ml of the aqueous extract is taken from a test tube to which a few drops of $FeCl_3$ at 1% are added. The appearance of a green coloration with or without precipitate indicates the presence of tannins.

6. Search for related quinones

To 1 ml of the aqueous extract add a few drops of Bornträger reagent (NaOH 10% or NH4OH 10%) and shake vigorously. The appearance of a colouring ranging from orange to bright red indicates the presence of bound quinones.

7. Check for alkaloids.

To 5 ml aqueous extract add 1 ml 1 N hydrochloric acid and 50µl Draggendorf's reagent, if an orange precipitate forms, this indicates the presence of alkaloids in the sample.

8. Search for saponosides.

The search for saponins is done by introducing a few milliliters of the aqueous extract into a test tube with a volume of distilled water and shaking for a few seconds. If the foam formed retains a height of at least 1 cm after 15 min of rest, the test is positive.

2.2.1.2. Preparation of the organic extract.

Macerate one gram of powder in 25 ml of dichloromethane for 24 hours with magnetic stirring, then filter.

> *Tests on the organic phase*

1. Testing for steroids and triterpenoids

Add Liebermann-Buchard reagent to the dry evaporated organic extract. A violet coloration indicates the presence of triterpenoids and steroids. Separately, the triterpenoids form a purple complex while the steroids develop a green coloration.

2. Search for free quinones

Free quinones are highlighted by treating the organic extract with Bornträger's reagent. The appearance of a colour ranging from orange to bright red is evidence of the presence of free quinones.

3. Search for triterpenoids

To the organic extract, add a few drops of concentrated H_2SO_4. A variation in coloration indicates the presence of triterpenoids.

4. Search for coumarins

The organic extract is treated with 10% NaOH. The presence of fluorescence under the U.V. lamp indicates that the test is positive.

2.2.2. Extraction by increasing polarity

Extracts with *n-hexane* (EH), dichloromethane (ED), ethyl acetate (EA) and methanol (EM) were obtained by successive extractions with solvents, according to the increasing order of their polarity. In this order, *n-hexane,* dichloromethane, ethyl acetate and methanol were used.

An appreciable amount of the powder of each plant is macerated successively in *n-hexane*, dichloromethane, ethyl acetate and methanol   for 48 h twice.

The extract is then concentrated in the Büchi Re 120 rotary evaporator and dried in the MEMMERT oven at 40°C.

The methanolic extract of *Ocimum gratissimum* was taken up again in distilled water and successively extracted by *n-hexane,* dichloromethane, ethyl acetate (liquid-liquid extraction). Then the ethyl acetate extract was subjected to column chromatography using Kieselgel 100 type silica gel of 0.2-0.5 mm/35 mesh for column.

Figure 2.1 below shows the extraction scheme with the solvents of increasing polarity used in this work.

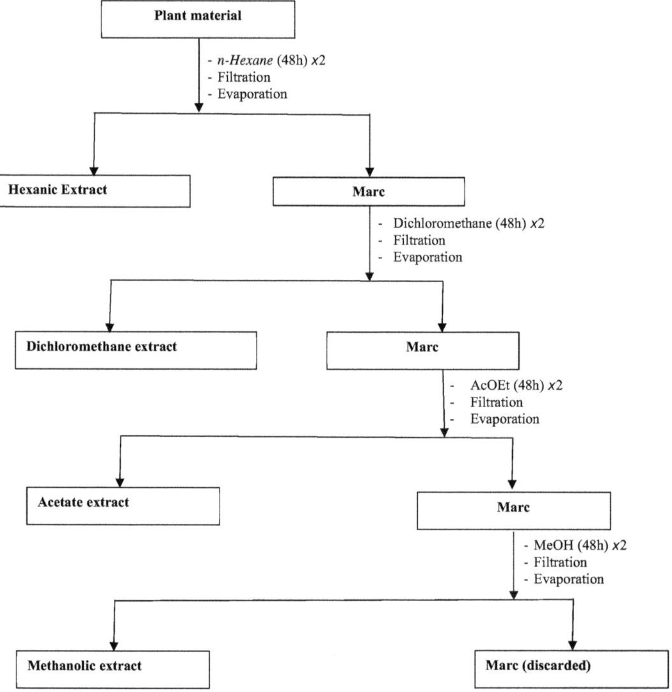

Figure 2.1: Extraction scheme with increasing polarity

2.2.3. Chromatographic screening and HPLC

2.2.3.1. Analysis by thin-layer chromatography

(a) Chemicals used

All solvents used were of analytical grade and originated from Merck VWR (Leuven, Belgium).

Rosmarinic acid, luteolin and vitexin were HPLC-type products supplied by "Sigma Aldrich". The water used was treated in the Easy pure purification system.

(b) Preparation of samples :

1 g of powder from each plant is extracted with 10 ml of methanol twice in an ultrasonic bath for 15 min and then filtered through Wathman filter paper.

One milligram of rosmarinic acid for HPLC analysis is dissolved in 10 mL methanol with gentle agitation. 10 µl of each extract are deposited in 1 cm strips on a silica gel chromatographic plate (Silicagel 60 F254) as a stationary phase and using as elution systems :

-   System I: ethyl acetate/formic acid/acetic acid/water (100:11:11:26)
-   System II: Formic acid/acetone/dichloromethane (8.5:25:85)

Rosmarinic acid, luteolin and vitexin were used as standards.

The plates were revealed by spraying them with natural product reagents (Neu Reagent/ PEG 400) and observed with a 365nm UV lamp. Flavonoids are detected by the presence of orange-yellow fluorescent spots and phenolic acids by blue fluorescent spots (Wagner, 1996).

2.2.3.2. Analysis by HPLC

Analyses are performed at 25°C using a "Argilent 1100" HPLC chain connected to the iodine strip detector. Extracts and standard products were filtered on Wathman filter paper. The filtrate is then filtered on a special 0.45 µm HPLC filter. And each sample is injected into the HPLC-UV/DAD system. Separation is performed using the HYPERSIL ODS column (4 mm x 250 mm) with a non-linear gradient of acetonitrile (solvent A) and 0.05% trifluoroacetic acid in ultrapure water (solvent B) in the proportions :

T0: 0 % A, 100 % B; T1: 3 % A, 97 % B; T45: 40 % A, 60 % B; T46: 0 % A, 100 % B and stopping at T60.

The time T is expressed in min. Compounds were eluted at a flow rate of 1mL/min and detected with Iodine Array Detectors (UV-DAD). UV spectra of elution peaks are drawn in the range 250-340 nm and chromatograms are viewed at 280 and 340 nm. Identification was based on retention time and absorption spectra compared to references and data available in the system database (Tsumbu *et al.*, 2011).

2.2.4. Separation and purification of phytochemical compounds

2.2.4.1. Thin layer chromatography

The analysis and separation of the extracts obtained by solvents with increasing polarity of plants into different constituents was carried out on a semi-rigid aluminium TLC plate of silica gel F 254, 0.02 mm thick (Kurt, 1971, Cherbuliez *et al.* , 1964).

The eluent (dichloromethane: DCM) and n-hexane-ethyl *acetate* (AcoEt); *n-hexane-dichloromethane* gradient elution systems were used. The developers used are the UV lamp (CAMAG type) at 366 and 254 nm and sulphuric acid 20% in MeOH, followed by heating at 100°C for a few minutes after spraying (Skoog et *al.*, 1997; Mahuzier & Hamon, 1998).

2.2.4.2. Column chromatography

The separation and isolation of a few components of some extracts were carried out on Kieselgel 100 type silica gel of 0.2-0.5 mm /35-70 mesh for column (Bourgeais and Florentin, 2002).

The monitoring of the composition of the fractions was carried out by CCM.

Column lengths, column diameters and the amount of silica gel used depended on the amount of extracts used and the separation of products on the chromatographic plate, with 30 to 50 g of gel being used for 1 g of extract per column (Horowitz, 2000).

Thus columns (L=30 cm, D=4 cm); (L=60 cm, D=2 cm); (L=40 cm D=3.7 cm), (L=15 cm, D=1 cm) were used with a polarity gradient: *n-hexane-DCM*, *n-hexane-AcoEt* and fractions of 20 mL were collected.

2.2.5. Elucidation of structures of isolated compounds

2.2.5.1. Nuclear Magnetic Resonance (NMR)

The NMR spectra were recorded using :

1) the Brüker Avance 300MHz spectrometer (for samples, Dor 2, Dor 4, Dina, Do8 and DM)
2) The JEOL JNM-LA400 spectrometer (for samples named Tsh-Do).

All spectra are recorded at laboratory temperature using deuterated chloroform and read from the deuterated chloroform reference line, which is δ_H =7.24 ppm for the [1H-NMR spectrum], and δ_C 77.20 ppm for the [13C-NMR] spectrum.

Samples were solubilized in the deuterated solvents $CDCl_3$ and CD_3OD in 5 mm diameter analytical tubes.

The chemical displacements (δ) are expressed in ppm relative to tetramethylsilane (TMS) while the coupling constants J are expressed in Hertz (Hz).

The standard pulse sequence programs provided by Brüker and Jeol -JNM LA allowed the two-dimensional experiments: COSY, DEPT, HMQC and HMBC to be carried out.

- **Homonuclear correlation**

- **COSY (COrrelated SpectroscopY)**: this experiment provides information on ^{2J} and ^{3J} homonuclear couplings (protons separated by two or three bonds).

- **Heteronuclear correlation**

- **DEPT (Distortionless Enhancement by Polarization Transfer)** which unequivocally specifies the number of CH, CH_2 and CH3 (Lanzetta et *al.*, 1984; Silvertein *et al.*, 2005)

- **HMQC (Heteronuclear Multiple Quantum Coherence)**: HMQC spectra indicate a correlation between a proton and the carbon to which it is fixed (highlights the ^{1J} coupling: this technique gives information on the DEPT- $^{1H\text{-}NMR}$ correlation (Silvertein *et al.*, 2005)

- **HMBC: (Heteronuclear Multiple-Bond Connectivity)** $(^{2JH\text{-}C}, {}^{3JH\text{-}C})$: this technique allows the detection of $^{2JH\text{-}C}$ and $^{3JH\text{-}C}$ long distance couplings. It is the long-distance correlation that makes it possible to see which proton is correlated with which carbon at more than one bond; it also makes it possible to deduce the quaternary carbons coupled to the protons (Silvertein *et al.*, 2005).

2.2.5.2. Mass spectrometry

Mass spectrometry of some compounds named Tsh-Do was performed using the JEOL JMS-700T LREIMS apparatus. The compound was dissolved in methanol (1:100) or 1g per 100 mL used for HPLC and the helium (0.8mL /min) used as elution gas was maintained at 200°C for 4 min and then programmed to $^{300°C}$ for 5°C per minute. Ionization is performed by electronic impact (EI) and the low and medium resolution analyzer used is time-of-flight (TOF).

2.2.6. Extraction of essential oils

The extraction of essential oils from the leaves of these three species of *Ocimum* is carried out by hydro distillation.

A known quantity of fresh material of each species, cut into small pieces, is placed in a 500 mL flask filled three-quarters full of water and brought to a boil. The distillate (water and gasoline) is collected in a separating funnel and separated by simple decanting; anhydrous magnesium sulphate is used to remove traces of water, and the oil, packaged in a brown bottle, is kept refrigerated at ±4°C (Roudmitska, 1980; Avlessi et *al.*, 2004, Guillaumin *et al.*, 1955; Loiseleur, 1963).

The operation is repeated several times to increase the amount of essential oil.

2.2.7. Analysis of essential oils

The chemical composition of the essential oils of *Ocimum gratissimum* and *O. basilicum* was determined by the gas chromatography (GC) technique and the gas chromatography/mass spectrometry (GC/MS) technique under the operating conditions described below.

2.2.7.1. Analysis by gas chromatography

The GC was done using a Hewlett-Packard chromatograph (HP 5880A series), equipped with a flame ionization detector (FID) and a capillary apolar column type OV-1 (30 m × 0.25 mm, film thickness: 0.25 μm). The carrier gas used is helium at a flow rate of 1 mL/min. The injection is done by split mode with a division ratio of 1/50. The volume injected is 1 μl. The column is initially maintained at a temperature of 40°C for the first five min, which is then programmed from 40-200°C at a rate of 4°C/min, then from 200-300°C at a rate of 10°C/min. The injector and detector temperatures were maintained at 250 and 280 °C respectively.

An internal calibration solution containing a mixture of *n-alkanes* (C9 to C20) is injected under the same conditions to calculate the retention indices of different peaks in the chromatograms of the essential oils analysed.

2.2.7.2. Analysis by gas chromatography/mass spectrometry (GC/MS)

The GC/MS was performed on a Varian CP 3800 chromatograph, equipped with an OV-1ms type apolar fused silica capillary column (30 m × 0.25 mm film thickness: 0.25 μm), coupled to a Kratos MS 80 RFA mass spectrometer (MS) equipped with a DS90 data system. The spectra were recorded between 50 and 240 amu. Fragmentation is carried out by electronic impact under a 70 eV field. The temperature of the transfer line is 295 °C. The chromatographic analysis conditions are the same as those described in the previous paragraph.

The identification of most of the constituents of these essential oils was made by comparing their mass spectra with spectra kept in different mass spectra libraries (NIST08,

Wiley Registry [10th]) and by comparing their retention indices with those known in the literature (El-Sayed, 2014; Adams, 2007).

For the essential oil of leaves and inflorescences of *Ocimum canum* Sims, the AGILENT type gas chromatograph and the gas chromatograph coupled to the SHIMADZU brand mass spectrometer were used. GC was performed on a chromatograph equipped with a 15m column containing (diphenyldimethylsiloxane 5%) with an FID detector. The mass spectra were recorded using a 30m capillary column Argilent DB-XLB chromatograph coupled to a Positive Electron Impact (EI) mass spectrometer. LRMS (low-resolution mass spectrum) data were obtained on a UPLC/MS-TOF apparatus equipped with an ESI interface and a C-18 column (1.7 μm, 2.1X50 mm) using ammonium carbonate (0.005 M)/methanol as the mobile phase.

2.2.8. Obtaining the aqueous and anthocyanin extracts of *Ocimum canum*

The aqueous extract of *Ocimum canum* is obtained by macerating the leaf powder in distilled water for 48 hours. The filtrate is oven-dried at 50°C; while the anthocyanic extract is obtained by macerating the leaf powder for 48 h in methanol acidified with 0.4M hydrochloric acid. Delipidation of the anthocyanin extract was carried out by liquid-liquid extraction with petroleum ether using a settling ball.

2.2.9. Synthesis of esters

The method used is esterification with sulphuric acid as a catalyst, taking into account the physico-chemical properties of the reagents and reaction products.

For all reactions, the fatty acids used are technical (about 85% acid).

In order to shift the equilibrium of the reaction, an initial excess of alcohol in the reaction medium (for reactions using fatty acids) and an excess of acid for reactions using simple acids with heavy alcohols (formic acid and acetic acid) were carried out; the water and any azeotrope formed was removed by continuous distillation using the Dean Stark apparatus.

Gradual heating of the reaction medium was carried out to speed up the reaction.

After esterification, the reaction medium is washed with water to remove residual alcohol, mineral acid (catalyst or H_2SO_4), simple organic acids (formic acid, acetic acid).

The separating funnel was used to separate the esters from the rest of the reaction medium and the traces of water were dried using anhydrous magnesium sulphate.

The products thus purified were analyzed by GC and GC/MS and identified by comparing their mass spectra with those of the Wiley database.

The general equation for the esterification reaction is given in Figure 2.2:

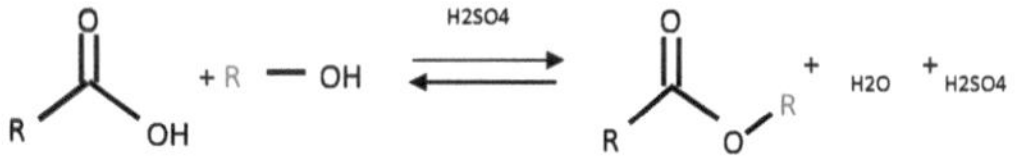

$$R-COOH + R-OH \underset{\longleftarrow}{\overset{H_2SO_4}{\longrightarrow}} R-COO-R + H_2O + H_2SO_4$$

Figure 2.2: General esterification equation.

2.2.10. Antifalcemic activity

The in vitro antifalcemic activity of different extracts, essential oils, some isolated (and standard) products and fatty acid esters has been demonstrated by the Emmel test (Yuma *et al.*, 2013).

The test consists of seeing the effect of the product on the falciformation of red blood cells. To perform it, a drop of the physiological solution is deposited on a slide, then the SS blood alone (control) or mixed with the total extracts or the different isolated products and finally everything is mixed with a drop of 2% sodium meta-bisulphite solution.

The object slide is covered with a cover sheet and the preparation is glued by means of the paraffin in supercooling deposited on the edge of the cover sheet.

The different preparations were examined with the Bresser Biolux NV optical microscope, at 5X and 16X magnification of the objective and coupled to the computer; images of red blood cells from different fields were captured with the webcam camera and processed with the software MOTIC IMAGES 2000 version 1.3.

2.2.11. Anti-radical activity

The assessment of 1,1-Diphenyl-2-picrylhydrazyl (DPPH°) scavenging was conducted as described by Philip (2004).

Briefly, a solution of the 0.3 mM DPPH radical in methanol was prepared by stirring with a Vortex Langdorp-Belgium CL001 Series 0902247 Vortex Blender. 3.5 mL of this solution is mixed with 0.5 mL of solutions of different concentrations of extracts (E), ascorbic acid (AA), essential oils (HE), and thymol, respectively.

All the solutions were prepared at various concentrations in methanol following progressive dilutions of geometric series of order 2, and depending on the case, some concentrations were retained: 3.93 to 1000 µg/mL (0.00393- 1.005 mg/mL) of the extracts, essential oil of *Ocimum gratissimum*, thymol and ascorbic acid with a few differences. For the essential oils of *Ocimum canum* and *O. basilicum* their concentrations ranged from 0.57 to 9.18mg/mL.

The mixture was primed and absorbances were read at a wavelength of 517 nm after each minute for 35 min considered as equilibrium time, using a UV-VIS 320 / SAFAS Monaco spectrophotometer. Ascorbic acid was used as a positive control and methanol as a blank control (Kallitraka et *al.* , 2005; Scherer *et al.* , 2009).

The percentage reduction of the DPPH radical is calculated from the relation

$$\% I = \frac{(A0-Aeq)}{A0} \times 100$$

Where A0 is the absorbance of DPPH° alone (negative control) and Aeq: the absorbance of the DPPH° mixture plus extracts, HE or pure products at equilibrium time.

The concentration of extracts or products at which 50% DPPH° is reduced in solution (IC50) or the dose required to inhibit DPPH radical formation by 50% was determined by linear regression from the % inhibition/concentration pair using Origin 6.1 software.

2.2.12. Antibacterial activity

The choice of bacterial strains used during this study is based for the most part on those developed by sickle cell disease patients, such as *salmonella, staphylococci, enterobacteria, Klesbsiella, Escherichia coli, streptococci, pneumococci* (Catonne *et al.*, 2012).

The bacterial strains were provided by the American Type Culture Collection (ATCC). They are maintained by transplantation on nutrient agar for 24 h, in the dark and at 37°C. They are :

- *Escherichia coli* ATCC 25922 (Gram-)
- *Staphylococus aureus* ATCC 25923 (Gram +)
- *Klesbiella pneumonia* ATCC 700603(Gram -)
- *Salmonella typhimurium* ATCC 14028 (Gram -)
- *Proteus mirabilis* ATCC 12453 (Gram -)
- *Pseudomonas aeruginosa* ATCC 2785 (Gram -)

Antibacterial activity is evaluated by the liquid micro-dilution method.

2.2.12.1. Preparation of the bacterial suspension

The bacterial suspension is prepared by placing 2 mL of physiological water in a sterile test tube; and using a sterile platinum loop, 3 colonies isolated from the test strain are placed in the physiological water in the tube. This bacterial suspension is diluted to one-tenth in TSB (Tryptic Soy Broth) culture medium. The resulting suspension has a turbidity of about 0.5 Macfarland (about 10^8 CFU/mL) (CFU: colony-forming units) (Strani *et al.*, 2007).

2.2.12.2. Dilution of essential oils and inoculation of the microplate

Sterile 96-well polystyrene microplates (8 rows A-H × 12 columns) are used. The essential oil is first emulsified with a drop of Tween 80 and then returned to the culture medium.

Procedure :

Fill each well with 50 µl of culture medium as follows: from A3-A11, B3-B11, C3-C11, D3-D11, E3-E11, F3-F11, G3-G11.

Using a 200 µl propette, 100 µl of the stock solution of the essential oils (20 µl in 1 mL of the medium) and the antibiotics (Cifen, Gentamycin and Augmentin) taken as positive controls, all at concentrations ranging from 0.98 to 250 µg/mL, i.e. 0.00098 to 0.25 mg/mL, are placed in wells B2, C2, D2, E2, F2, G2 respectively. Take 50 µl of the stock solution of the essential oil or the positive control in well No. 2 and make dilutions in geometric series of order 2. Mix well (by aspiration and successive discharges into the propette) as follows:

Take 50 µL from well #2 and transfer it to well #3. Mix well and continue with the same procedure to well #9. The last 50 µl taken from this last well will be discarded.

Similarly, aseptically withdraw 50 µl of the standardised inoculum with a micropipette, transfer to all wells of the microplate except well No. 11 (culture medium sterility control) and well No. 12. Well #10 was used as a growth control (inoculum and culture medium). Incubate the microplates in an oven at 37°C for 24 hours.

2.2.12.3. Determination of the minimum inhibitory concentration Inhibitor

The Minimum Inhibitory Concentration (MIC) is defined as the lowest concentration of the essential oil that completely inhibits bacterial growth in the wells of the microplate; an inhibition detectable by the naked eye.

The growth in the wells containing the EO or control should be compared to the growth in the bacterial growth control wells (wells with inoculum without EO or antibiotic). For a test to be considered valid, acceptable growth in the growth control wells must be observed. If there is insufficient growth in these wells, reincubate the microplate and read the MIC after 48 hours. The MIC will be read after 20 µL of resazurin 2%, a life indicator, has been added to the test wells.

The principle of this method is based on the ability of living cells to reduce blue resazurin to pink resofurin. On reading after the addition of 20 µl of resazurin 2% to each test well, the pink colour indicates bacterial multiplication; the persistence of the initial blue colour indicates the absence of germ growth. The MIC is the lowest concentration at which no visible growth has been noted. It is then read from the first wells with no bacterial growth.

2.2.12.4. Determination of minimum bactericidal concentration (MBC) of essential oils

The BMC is determined by seeding on TSA (Tryptic Soy Agar) medium, starting from the MIC, with all wells showing no visible bacterial growth. In this way, the bactericidal or bacteriostatic concentrations of our essential oils are determined.

Briefly, the contents of each well, ranging from the MIC value to the highest concentrations, were streaked on the surface of TSA nutrient agar. The wells of the positive and negative control rows were also streaked on the same agar to ensure that there was no growth of bacteria in the wells of the negative control row and no bacteria in the wells of the positive row. The inoculated plates were incubated at 37°C for 24 hours. On reading, the BMC is the lowest concentration of HE for which there was an absence of surviving bacteria in the test wells.

CHAPTER III: RESULTS AND DISCUSSION

3.1. Phytochemistry

3.1.1. Chemical screening

Table 3.1. below gives the results of the chemical screening carried out on the leaves of *Ocimum basilicum, O. canum and O. gratissimum.*

Table 3.1: Result of the chemical screening of aqueous and organic extracts of three species of the genus *Ocimum.*

Natural substances	*O.basilicum*	*O.canum*	*O.gratissimum*
1. Polyphenols	+	+	+++
- Flavonoids	+	+	+
- Anthocyanins	+	+	++
- Leucoanthocyans	+	+	++
- Tannins	+	+	+++
- Linked Quinones	+	+	++
2. Alkaloids	+	+	+
3. Saponines	+	-	+++
4. Triterpenoids	+	+	+
5. Steroids	+	+	+
6. Free quinones	+	+	+

Legend :

+++ : Very abundant presence of the desired substance

++ : Abundant presence of the substance sought

+ : presence of the substance sought

- absence of the substance sought

From this table 3.1 it can be seen that the aqueous and organic extracts of the leaves of these three plants : *Ocimum basilicum, O. canum* and *O. gratissimum,* contain polyphenols (flavonoids, anthocyanins, leucoanthocyans, tannins, quinones), alkaloids, triterpenoids and steroids.

The moss test demonstrated the presence of saponins in both species: *Ocimum basilicum* and *O. gratissimum* and this group of natural substances is absent in *Ocimum canum.*

Polyphenols appear to be more abundant in *Ocimum gratissimum* than the other two species. This is confirmed by the abundant presence of anthocyanins, leuco-anthocyanins, tannins and quinones in the same species. Sensitivity to the flavonoid test appears to be the same for all three species.

This result was confirmed for *Ocimum gratissimum in* a spectrophotometric assay study on the species harvested in Côte d'Ivoire, which had a higher content of polyphenols and flavonoids than other species of other genera (N'Guessan *et al.*, 2011).

The presence of high polyphenol content has been reported for all three species harvested elsewhere (Ganiyu, 2008, Jamal *et al.*, 2002).

The foaming power test, confirming the presence of saponins, is more sensitive in *Ocimum gratissimum* followed by *O. basilicum* and less sensitive or negative for *O. canum*.

The results obtained indicate approximately the same content of free alkaloids, triterpenoids, steroids and quinones for the three species.

These chemical screening results therefore show that two of the three *Ocimum* species (*O. basilicum* and *O. canum*) have the same phytochemical groups in approximately equal content except for saponins. Saponins are non-existent in O. *canum*.

This preliminary analysis, which aims to know the composition of secondary metabolites of these three plants, shows that they contain all the chemical groups tested (*Ocimum basilicum* and *O. gratissimum*), except the saponins which are absent in *Ocimum canum*.

3.1.2. Fractional extraction

Table 3.2. below groups the different extracts obtained with 4 solvents of increasing polarity and their percentage yield for each species.

Table 3.2: Extraction yield for each plant

Plant name	Solvent/Extraction yield in % Solvent			
	n-Hexane	Dichloromethane	Ethyl acetate	Methanol
O. basilicum	2,17	1,79	1,27	1,87
O. canum	0,59	0,87	1,86	4,09
O. gratissimum	1,76	2,67	1,26	3,31

From Table 3.2 it can be seen that each plant has a particular behaviour:
The yield of hexanic extract is higher than that of three other extracts of *Ocimum basilicum,* but the difference is not very large and the yield of DCM extract is almost the same as that of the MeOH extract of the same species (1.79 and 1.87 % respectively).

The yield of different extracts of *Ocimum canum* is directly proportional to the polarity, the higher the polarity the greater the yield (from 0.59 to 4.09%). The same is true for *Ocimum gratissimum,* except that the yield of the extract in DCM is higher than that of the ethyl acetate extract.

For all three species: the yield of *n-hexanic* extract of *O. basilicum* is higher than those of two other species; the yield of dichloromethane extract of *O. gratissimum is higher than those of two other species*; the yields of ethyl acetate and methanolic extract of *O. canum are* higher than those of two other species.

3.1.3. Screening by CCM and HPLC

3.1.3.1. Analysis by Thin Layer Chromatography

Figures 3.1. (a) and (b) show the TLC chromatograms of methanolic extracts of three species of the genus *Ocimum* first alone, and then with phenolic controls respectively.

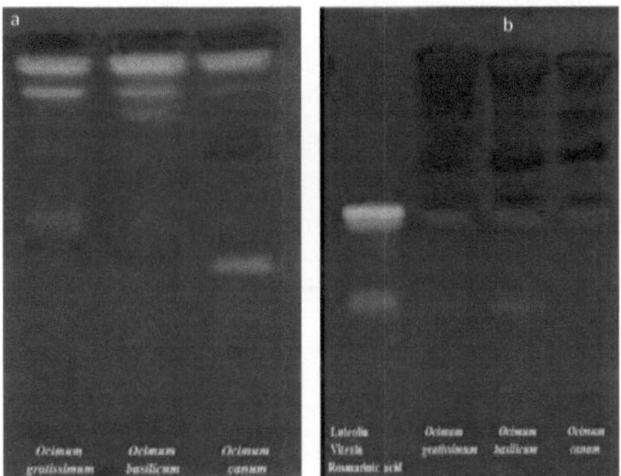

Figure 3.1. The TLC chrommatogram of methanolic extracts from *O. basilicum*, *O. canum* and *O. gratissimum*; developed with the system Ethyl acetate/formic acid/ glacial acetic acid/water 100:11:11:26 (a) and the system: dichloromethane/acetone/formic acid 85:25:8.5 with rosmarinic acid, luteolin and vitexin as standards (b) visualized at 365 nm with the PEG reagent of Products (Neu) .

Flavonoids are detected by the presence of yellow-orange fluorescent spots and phenolic acids by the presence of blue fuorescent spots.

As indicated; chromatographic analyses by TLC show that the methanolic extracts of these three *Ocimums* contain the polyphenols. Indeed, considering chromatogram(a), the presence of flavonoids of the Kämpferol type (the yellow-orange fluorescent spots) and phenolic acids (the blue fluorescent spots) is noted; and considering chromatogram(b), rosmarinic acid is identified by coincidence of the Rf of the control with the correspondents in the three extracts.

The presence of rosmarinic acid is confirmed by superimposing HPLC chromatograms of the pure rosmarinic acid taken as a control, and that of the methanolic extract of each of three *Ocimum*.

3.1.3.2. Analyses by high-performance liquid chromatography (HPLC)

Figure 3.2 shows the structure and HPLC profile of pure rosmarinic acid (RA) as a control for comparison of HPLC profiles of methanolic extracts from the three *Ocimum*

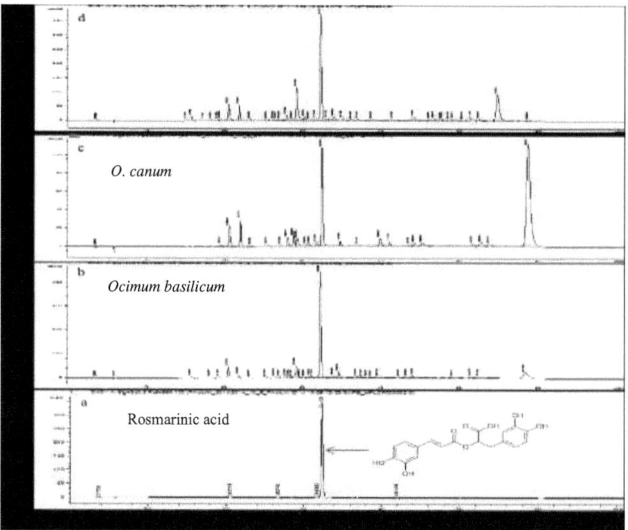

Figure 3.2: HPLC-DAD chromatograms of rosmarinic acid (a) and methanolic extracts of *O. basilicum* (b), O. *canum* (c), O. *gratissimum* (d) using a non-linear gradient of acetonitrile (A) and trifluoroacetic acid 0.05% in ultrapure water (B): 0 min, 0:100 (A:B), 1 min, 3:97(A:B); 45 min, 40:60 (A:B); 46 min, 0:100 (A:B) 60 min, stopped at a flow rate of 1mL/min on a Hypersil ODS column (4 mm × 250 mm) with detection at 340 nm.

The elution time is given in min (horizontal axis). On the Y-axis, each elution peak expressed in arbitrary milli units (mAU) corresponds to the concentration of the eluted compound in the test sample.

As can be seen in Figure 3.2, pure rosmarinic acid AR (a) appears as a single peak at the retention time of 32.408 min. As noted above, the methanolic extract profile of *O. canum* (c) shows two intense peaks. The most intense peak and therefore the majority compound appears at a retention time of 58.719 min, followed by the peak at 32.494 min. The latter is approximately equal to the retention time of AR.

These results are close to those found by Berhow *et al* (2012) who found a retention time of 30.07 min. The presence of AR in the methanolic extract of *O.canum* is confirmed by the coincidence of two peaks: that of AR and that characteristic of the profile of the methanolic extract at the same retention time. The data available in the database used as references for the system did not allow us to identify the compound with a retention time of 58,719 min.

We can also see on the HPLC profile of *O.basilicum* (b) that the most intense peak is at the retention time equal to 32,300 nmin. This value is approximately equal to that of AR. These results are identical to those found for the methanolic extract of *O. basilicum from* Romania (Vlase *et al.,* 2014).

The presence of RA is demonstrated by superimposing the peak of the HPLC profile of pure rosmarinic acid with the corresponding peak of the HPLC profile of the methanolic extract of *O. basilicum* at the same retention time.
When comparing the two peak intensities identified in the RA for *Ocimum canum* and *O. basilicum*, the peak of the latter (b) appears to be more intense than the former (c).
Thus, this result confirms that *O. basilicum* contains more AR than *O. canum.*

It is also noticed on the same figure 3.2, for the HPLC profile of *Ocimum gratissimum* (d), the presence of the most intense peak at retention time 32,310 min which is close to that of rosmarinic acid.

This result is comparable to that found for the methanolic extract of *Ocimum gratissimum from* Brazil (Costa *et al.,* 2012).

As for the first two methanolic extracts, the presence of rosmarinic acid is demonstrated by the superposition of two peaks of pure RA and that of the methanolic extract of *O. gratissimum* at almost the same retention time.

Comparing the three HPLC profiles of the methanolic extracts of the three species taking into account their peak intensities, *O. basilicum* contains a higher content of rosmarinic acid while *O. canum* contains less than the other two.

3.1.4. Analytical chromatography of extracts of *Ocimum gratissimum*

3.1.4.1. Thin layer analytical chromatography

The different constituents in the hexane, dichloromethane (DCM), ethyl acetate (AcOEt) and methanolic (MeOH) extracts of *Ocimum gratissimum* were detected using the eluents and elution systems corresponding to the same solvents, i.e. *n-hexane*, DCM, AcOEt or binary mixtures.

3.1.4.2. Column chromatography

The three extracts DCM, AcOEt and methanol recovered in the AcOEt after washing with water and liquid-liquid extraction with *n-hexane* and dichloromethane for *Ocimum gratissimum*, were used to separate on column some constituents revealed by CCM.

a) Column of the extract with DCM of Ocimum gratisimum

Column chromatography of the dichloromethane extract was started by successively eluting with *n-hexane* alone: 200mL (10 fractions of 20mL each); then with each 200mL of the *n-hexane* - DCM increasing polarity gradient system (9-1); (8-2); (7-3); (6-4); (5-5); (4-6); (3-7) : (2-8). After CCM monitoring, seven fractions were grouped together and named F1, F2, F3, F4, F5, F6, F7. And F8 is the one obtained after washing the rest of the column with ethyl acetate.

The different fractions from this first column of the dichloromethane extract, the eluent and the elution systems used are shown in the following Table 3.3:

Table 3.3: Different fractions from the DCM (ED) extract column; eluent and elution systems used

Glass no. (20mL)	Eluent or eluent system	Fractions

1-3	n-hexane	F1 : mixture of 2 products at UV254 and H_2SO_4
4-9	n-hexane	F2
12-20	*n-hexane-DCM (9-1)*	F3
21-45	*n-hexane-DCM (8-2)*	F4
46-69	*n-hexane-DCM (7-3)*	F5
70-154	*n-hexane-DCM (4-6)*	F6
155-193	*n-hexane-DCM (2-8)*	F7

After cleaning the column with ethyl acetate, the large fraction called **F8 or F8/ED** was collected. Since the quantity of this last fraction is sufficient to continue its analysis, its TLC performed with the *n-hexane-AcoEt* system (7-3) in the 3rd position from the head of the plate shows a pink product which appears after revelation with H_2SO_4 followed by heating of the plate and yellow with the UV-366nm lamp which testifies that it would be a triterpenoid (Ndom, 2008).

The F8 column was eluted successively with *n-hexane* alone (3 glasses) and then each time with 200mL of the *n-hexane-AcoEt* polarity gradient system: 9.5-0.5; 9-1; 8.5-1.5; 8-2; 3.5-2.5; 7-3(4X); 6-4(2X). TLC monitoring gives from the 67th glass to the 76th a pink product after revelation at H_2SO_4 and yellow at UV 366nm with a small pink contaminant appearing at UV 366nm which has been purified by passing it over the activated carbon.

Table **3.4** below lists the various small fractions and products from the large F8 fraction of the DCM extract, the eluent and the eluent systems used.

Table 3.4: Different small fractions and product from the F8 fraction of the DCM extract (F8/ED); eluent, eluent systems used

N° of glasses	Eluent or eluent system	Separate product and fractions
1-3	*n-hexane*	F1
4-14	n-hexane-AcoEt (9.5-0.5)	F2
15-24	*n-hexane- AcoEt (9-1)*	F3
25-37	*n-hexane- AcoEt (8.5-1.5)*	F4
38-46	*n-hexane- AcoEt (8-2)*	
46-49	*n-hexane- AcoEt (3.5-2.5)*	F5
50-56	*n-hexane- AcoEt (7-3)*	

57-67	*n-hexane*- AcoEt (7-3)	
67-76	*n-hexane*- AcoEt (7-3)	**Dor 4**
80-100	*n-hexane*- AcoEt (7-3)	F6
101-126	*n-hexane*- AcoEt (7-3)	F7

In order to isolate a large number of compounds from the extract at the DCM displaying several spots on the TLC plate, and which is in large quantity (26.66g), the column was reworked twice with this extract using fractionated elution in a stepwise manner without abruptly changing the polarity by joining to *n-hexane* :

1) Dichloromethane: the products or samples obtained are named Dorl, Dor 2, Tsh-Do/2/II (Table 3.5).

2) Ethyl acetate: the samples obtained are called Dina (Table 3.6).

Table 3.5. Different fractions from ED extract; eluent used (*n-hexane* followed by DCM) and separate products

N° of glasses	Eluent or eluent system	Separate products and fractions
1-3	*n-hexane*	Dorl
4-68	*n-hexane*	**Dor 2**
69-100	*n-hexane*-DCM (9.5-0.5)	Dor3 blend
82-96		F1
100-160		F2
170-200		Tsh-Do1/2/II
201-240		F3
241-479		Tsh-Do2/2/IIS (appears at $_{H2SO4}$)
480-499		F4
500-532		Tsh-Do3/2/II
572-629		F5
630-639		F6
640-682		F7

Table 3.6: Different fractions and products from the ED extract; eluent used (*n-hexane* followed by AcOEt

N° of glasses	Eluent or eluent system	Separate products and fractions
1-13	*n-hexane*	Dinal (after $_{H2SO4}$)
14-30	n-hexane	Dina 2 (red liquid)
31-40	*n-hexane-AcoEt* (9.5-0.5)	F1 (UV254 and pink to $_{H2SO4}$)
45-89		F2
90-120		F3
125-140		F4 (UV254; Rf=0.5)

141-159		F5
160-235		Mix 3 products with H2SO4 F6
241-262		F7
263-312		F8
317-332		F9
630-639		F10
640-682		F11

Products or samples Dina3, Dina4, and Dina5 are the precipitates observed by recovering fractions F1, F2, and F3 with ethyl acetate.

b) *Column of the ethyl acetate extract of Ocimum gratissimum*

For this extract, 10 samples of 20mL each were taken using *n-hexane* alone and then the n-hexane-AcOEt eluent system (6-4) followed by the *n-hexane-AcOEt* system (5-5). The fractions resulting from this operation and the eluent systems used are grouped in Table 3.7. below.

Table 3.7: Different fractions from the EA extract column; eluent and eluent systems used.

N° of glasses	Eluent or eluent system	Fractions
1-10	*n-hexane*	
11-24	*n-hexane* -AcOEt (6-4)	
25-60	*n-hexane-AcOEt* (5-5)	F1
100-235	*n-hexane-AcOEt* (5-5)	F2
	MeOH	F3

- **F2 and F3 column**

Fractions F2 and F3, which had the same TLC profile, were mixed and fractionated in the column using DCM to isolate a few compounds named Tsh-Do/3/II. **F3** is the large fraction resulting from washing the column with methanol. Thus, the different fractions and the eluent are listed in Table 3.8 below:

Table 3.8: Different fractions and products from the fraction (F2, 3) of EA; eluent used and products separated

N° of glasses	Elective	Separate products and fractions
241-281	DCM	Dor6

285-300	DCM	1product at 366 nm blue Dor7
301-340	DCM	F1
341-351	DCM	F2
352-386	DCM	Tsh-Do1/3/II (pink product after H_2SO_4)
387-400	DCM	F3
410-420	DCM	Tsh-Do2/3/II
421-439	DCM	Mixture
440-490	DCM	Product blue 366Tsh-Do 3/3
510-540	DCM	Tsh-Do4/3/II
550-571	DCM	F4
572-683	DCM	F5
684-709	DCM	Tsh-Do5/3/II(visible after H_2SO_4)
710-739	DCM	Tsh-Do5/3/II
764-779	DCM	Tsh-Do5/3/II with1small contaminant
781-830	DCM	Tsh-Do6/3/II: yellow after H_2SO_4
831-880	DCM	F6
900-920		1prepity using MeOH

F1 to F6 are the different fractions received.

c) Methanolic Extract Column

The methanolic extract was washed in water and then recovered successively in *n-hexane*, DCM and ethyl acetate by liquid-liquid extraction.

The products obtained from the extract recovered from Ethyl Acetate by fractional elution using *n-hexane and* then the *n-hexane-AcOEt* mixture are referred to as DM. The different fractions and isolated products of the methanolic extract recovered in AcOEt and the elution systems used are grouped in the following Table 3.9.

Table 3.9: Different fractions and samples from the EM(AcoEt) column; eluent and eluent systems used

Glass no. (20mL)	Elective	Separate products or
1-40	*n-hexane*	F1 yellow: Precipitated
41-59	*n-hexane*	F2
60-85	*n-hexane*	DM1 to Rf=0.925 precipitate
90-130	*n-hexane-AcOEt* (9.5-0.5)	F3 at 254nm Rf=0.6
131-148	*n-hexane-AcOEt* (9.5-0.5)	
170-185	*n-hexane-AcOEt* (9.5-0.5)	DM2: red liquid
189-230	*n-hexane-AcOEt* (9-1)	F4
235-260	*n-hexane-AcOEt* (9-1)	F5

270-310	*n-hexane-AcOEt* (9-1)	DM3 precipitated black
320-370	*n-hexane-AcOEt* (9-1)	F6
372-390	*n-hexane-AcOEt* (9-1)	F7
391-420	*n-hexane-AcOEt* (9-1)	F8
430-480	*n-hexane-AcOEt* (9-1)	DM4
481-520	*n-hexane-AcOEt* (9-1)	DM5: at 254 Rf=0.525
530-572	*n-hexane-AcOEt* (9-1)	F9
586-648	*n-hexane-AcOEt* (8.5-1.5)	DM6: Precipitated black
649-681	*n-hexane-AcOEt* (8.5-1.5)	F10: 4 products
686-741	*n-hexane-AcOEt* (8.5-1.5)	F11
900-920		DM7

F1 to F11 are the fractions obtained (mixture of several products)

Since it was believed that all samples not labelled F were pure or less contaminated products, spectroscopic and chromatographic analyses were carried out on them to determine their chemical structures.

3.1.5. Analysis by spectroscopy and gas chromatography of samples (or products) isolated from *O. gratissimum*

3. 1.5.1. Analysis of Dina samples from the extract at the DCM (ED)

a) Analysis of the Dina1 sample

Compounds from the **Dina1** sample were isolated in a mixture as a whitish precipitate from the DCM extract.

The [1H-NMR] and [13C-NMR] spectra (Figures 4.1 and 4.2 in the Appendix) recorded in CDCl3 show that the Dina1 sample is a mixture of products. The mixture comprises a major aliphatic hydrocarbon and a compound or compounds containing carbon-carbon double bonds.

1) An alkane: Data from NMR spectra suggest that the predominant product is likely ***n-tetradecane*** (Pouchert & Behnke, 1993).

[1H-NMR]δ_H (ppm):   0.87 (bs)   1.   26(s)   0.87(bs)

CH3 - CH2 - CH2 - CH2 - CH2-(CH2)6- CH2 - CH2 - CH2 - CH2 - CH2- CH3

[13C-NMR]δ_C (ppm):   14.1   22.7   31.9   29.5   29.7   29.7 29.5   31.9   22.7 14.7 14.1

2) Derivatives of unsaturated fatty acids and aromatic hydrocarbons: minority.

This is confirmed by GPC and GC/MS analysis which, by comparing the spectra of some Dina1 compounds with the spectra from the database in the machine (Wiley 275 L) which prove the presence of *n-tetradecane* at the retention time 10,741min and other unsaturated compounds.

The chromatogram of the **Dina1** sample (**Figure 3.3.**), indicates that the injected sample is a mixture of at least 26 compounds by obtaining at least 26 peaks and of which those between 10 and 13 min as retention time are the majority and are identified by comparison of their mass spectra with those of the database.

Thus the peak (11,254 min) corresponds to the trans-karyophyllene

The peak (11,833 min) corresponds to β-bisabolene

The peak (12,113 min) corresponds to the epi-α selinene.

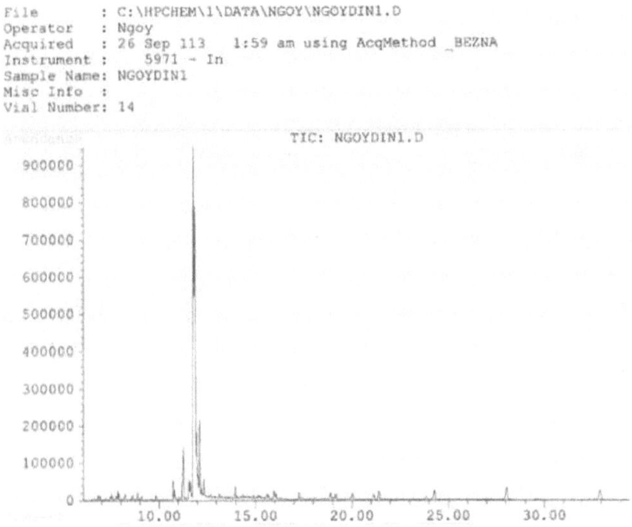
File : C:\HPCHEM\1\DATA\NGOY\NGOYDIN1.D
Operator : Ngoy
Acquired : 26 Sep 113 1:59 am using AcqMethod _BEZNA
Instrument : 5971 - In
Sample Name: NGOYDIN1
Misc Info :
Vial Number: 14
TIC: NGOYDIN1.D
900000
800000
700000
600000
500000
400000
300000
200000
100000
0
10.00
15.00
20.00
25.00
30.00

Figure 3.3: Chromatogram of the Dina1 sample

b) Analysis of the Dina2 sample

The Dina2 sample was isolated as a red liquid from the DCM extract using *n-hexane* as eluent.

The analysis of the $^{1H\text{-}NMR}$ spectrum recorded in the CDCl3 allowed to observe the presence of only three signals between 2.15 and 1.18ppm (Figure 4.3 see Appendix 1).

The analysis of $^{13C\text{-}NMR}$ spectra (Figure 4.4 see Appendix 1) shows a single peak at 30.9 ppm. The chemical displacements, multiplicities and corresponding groups are shown in Table **3.10** below.

Table 3.10.: ^{1H} Spectrum - Dina 2 NMR

δ_H (ppm)	Multiplicities	Groups

1.19	*s*	
1.49	*s*	
2.10	*s*	$-(CH_2)_n-$

The peaks at δ=1.19 and δ=1.49 would be impurities.

The [13C-NMR] spectrum shows only a single peak at 30.9 ppm, corresponding to magnetically equivalent methylene groups : $-(CH_2)_n-$. This suggests that the majority product would be a cycloalkane (cyclohexatetracontane).

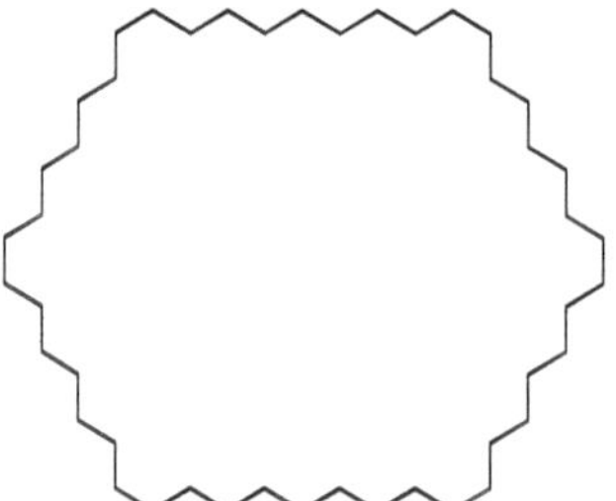

Figure 3.4: Structure of cyclohexatetracontane

GIC/MS analysis was performed on the Dina2 sample.

The Dina2 chromatogram indicates that the injected sample is a mixture of approximately six compounds with six visible peaks (**Figure 3.5**).

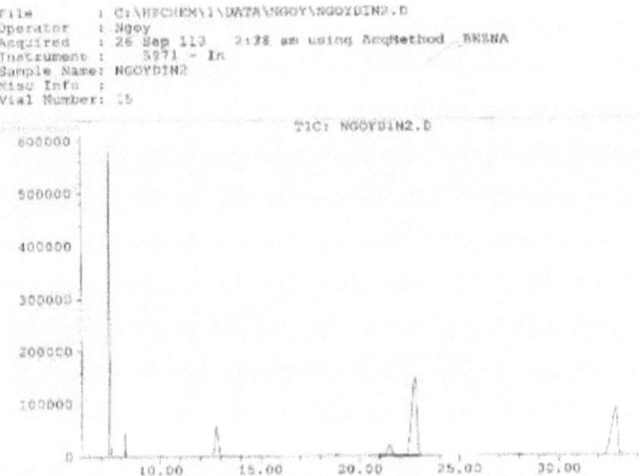

Figure 3.5: Chromatogram of Dina2

The mass spectra recorded for five compounds are compared with their counterparts from the instrument's Willey 275 L database. Thus, these five compounds are identified as diethyl dithiocarbonate (8,242 min), phenicosan (12,816 min), nonadecane (21,487 min), (6E,10E,14E,18E)-2,6,10,15,19,23-hexamethyltetracosahexa-2,6,10,14,18,22-ene or squalene (22,759 min) and nonacosane (32,887 min).

The compound with a retention time equal to 7.5min would be cyclohexatetracarontane projected by [1H-NMR] and [13C-NMR] which is the most abundant given the intensity of its peak.

c) Analysis of the Dina3 or DN3 sample

Based on GPC/MS data from DN.3D or Dina3, this fraction is a mixture of at least 10 products (Figure 3.6).

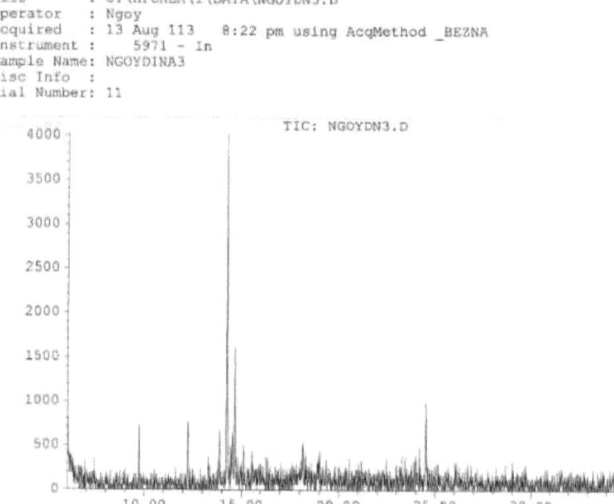

Figure 3.6: Chromatogram of Dina3

NMR data also confirm that Dina 3 is a mixture of products. The [1H] and [13C NMR] spectra of Dina3 (Tables 3.11 ,3.12 and Figure 4.5. in the Appendix) indicate the presence of a majority aliphatic hydrocarbon, which would be heptadecane (Charles *et al.,* 1993).

Table 3.12 also indicates the presence of two OH-group bearing carbons.

Table 3.11.: $^{1H\text{-}NMR}$ Spectrum of Dina 3

δH (ppm)	Multiplicities	Groups
0.84	*s*	-CH3
1.26	*s*	-(CH2)n-
1.51	*s*	-CH2-

Table 3.12.: $^{13C\text{-}NMR}$ Spectrum of Dina 3

Woodpecker	δc (ppm)	Group
1	14.1	*CH3-* (terminal)
2	22.7	-CH2 - CH3
3	29.5	*-CH2-*
4	29.7	*-CH2-*
5	31.9	*-CH2-*
6	62.1	*-CH-O-*
7	64.9	*-CH-O-*

Mass spectra data for two minority products of Dina3 obtained by GC/MS are shown in **Tables 3.13 and 3.14**. They likely correspond to *(Z)*-**5-undecen-2-ol** and *(E)*-**5-undecen-2-ol** (Beauchamp, 2013).

Figure 3.12 below explains some of the fragments observed for the compound *(Z)-5-undecen-2-ol*:

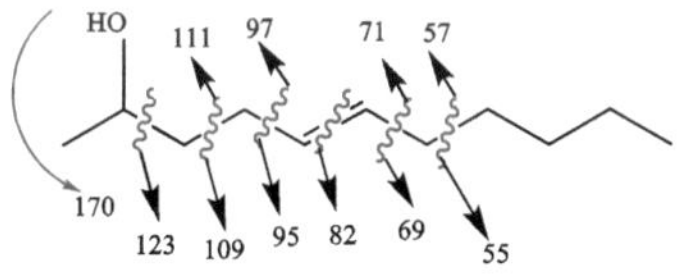

Figure 3.7: Fragmentation of *(Z)-5-undecen-2-ol* in MS

Table 3.13.: DN3.D Mass Spectrum Data (Scan 733:14,244 min.) : *(Z)-5-undecen-2-ol*

Woodpecker	m/z	Intensity (%)	Fragments
1	170	13.1	M+
2	124	25.3	M-46 (C_2H_5OH+)
3	123	38.9	M-47 (C2H5OH2+)
4	111	13.3	M-59 (C_3H_6OH+)
5	109	30.6	123-14 (-CH2-)
6	97	40.4	111-14 (-CH2-)
7	96	22.2	111-14, -H+ (-CH2-, -H+)
8	95	84.4	109-14 (-CH2-)
9	83	13.8	97-14 (-CH2-)
10	82	100	95-13(CH=)
11	81	48.9	95-14 (-CH2-)
12	79	13.8	
13	71	33.8	C4H9+ (CH3- CH2-CH2-CH2-)$^+$
14	70	21.1	
15	69	72.9	82-13(CH=)
16	68	68.9	
17	67	52.2	
18	65	13.3	
19	57	81.1	C4H9+ (CH3-CH2-CH2-+)
20	55	53.3	69-14 (-CH2-)
21	53	18.2	
22	52	12.9	

The second spectrum shows the same fragmentation profile, suggesting that this product is an isomer of the previous product. The same base peak and most of the fragments formed in the mass spectrum of *(Z)-5-undecen-2-ol* are observed.

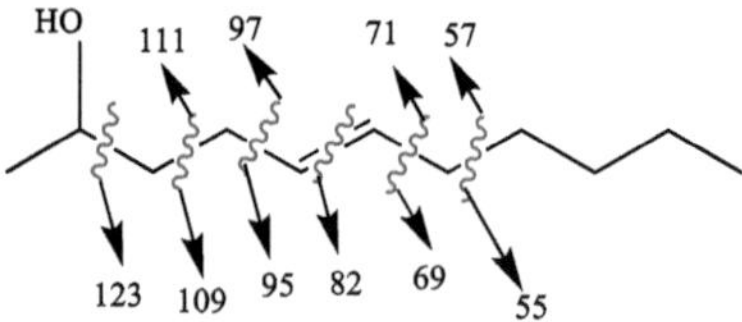

Figure 3.8: Fragmentation of *(E)-5-undecen-2-ol* in MS
Table 3.14. Dina3.D Mass Spectrum Data (Scan 769) : *(E)-5-undecen-2-ol*

Woodpecker	m/z	Intensity (%)	Fragments
1.	123	18.5	M-47 (C2H5OH2+)
2.	111	19.3	M-59 (C3H7O+): C8H15+.
3.	97	24.1	111-14 (-CH2-): C7H13+.
4.	96	44.4	M-74 (C4H10O+)
5.	95	24.4	123-28 (-CH2-CH2-)
6.	83	25.9	97-14 (-CH2-): C6H11+.
7.	82	100	96-14 (-CH2-)
8.	81	20.4	95-14 (-CH2-)
9.	79	33.3	83-14 (-CH2-)
10.	77	21.5	81-14 (-CH2-)
11.	71	23.0	
12.	70	31.5	
13.	69	45.2	55+14 (-CH2-): C5H9+ (CH3-CH2-CH2-
14.	68	24.1	82-14 (-CH2-)
15.	67	23.0	81-14 (-CH2-)
16.	57	73.8	71-14 (-CH2-): C4H9+ (CH3-CH2-CH2-+)
17.	55	33.3	69-14 (-CH2-): C4H7+ (CH3-CH2-CH2-+)

Although the two compounds have similar spectra, the absence of the molecular ion peak and the peak at m/z = 85 in the (E)-isomer, and the difference in peak intensity at m/z = 83 allowed the two isomers to be distinguished (Beauchamp, 2013).

The presence of two weak peaks at δ62.1 ppm and δ64.9 ppm on the [13C-NMR] spectrum of Dina3 corresponding to the oxygen bearing carbons (C-OH), suggests the presence of these compounds in the DINA3 mixture although the peaks of the olefinic carbons are not observed. In addition, [1H] and [13C-NMR] data indicate that the majority product would be an aliphatic hydrocarbon, **heptadecane** (Charles *et al.,* 1993).

The [1H-NMR] data also indicate that this sample is contaminated with aromatic compound(s).

d) Analysis of the Dina4 sample

The Dina 4 fraction has almost the same $^{1H\text{-}NMR}$ profile as Dina 3. It indicates the presence of a majority aliphatic hydrocarbon confirmed by the ^{1H} and $^{3C\text{-}NMR}$ data (**Figures 4.9 and 4.10 see appendices**).

On the other hand, the mass spectrum of the majority product obtained by GC/MS (Figure 4.11 see Appendices) does not correspond to an aliphatic hydrocarbon. Rather, it would be close to an aromatic hydrocarbon, in this case phenyl-2-tipyl-2-acenapthenone or its isomer (NIST Chemistry Web Book).

It could also be a phthalate diester, contaminating certain solvents. The mass spec profile suggests that.

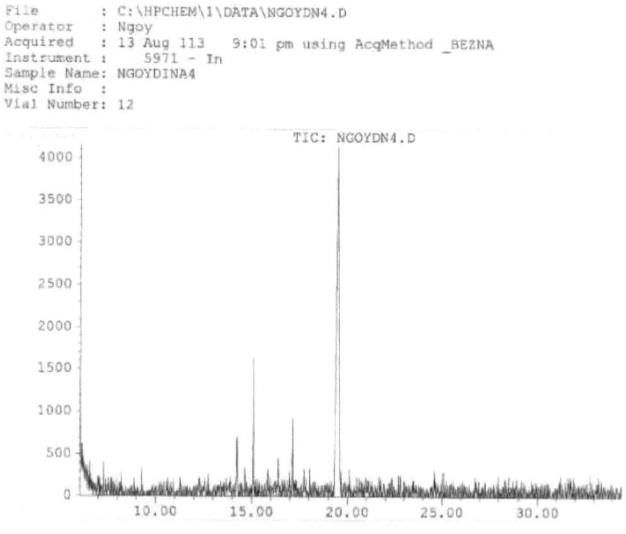
File : C:\HPCHEM\1\DATA\NGOYDN4.D
Operator : Ngoy
Acquired : 13 Aug 113 9:01 pm using AcqMethod _BEZNA
Instrument : 5971 - In
Sample Name: NGOYDINA4
Misc Info :
Vial Number: 12
TIC: NGOYDN4.D
4000
3500
3000
2500
2000
1500
1000
500
0
10.00
15.00
20.00
25.00
30.00

Figure 3.9: Chromatogram of Dina4

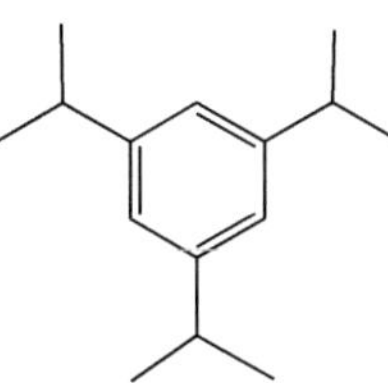

Phényl-2-tipyl-2-acénapthénone

Tipyl= 2,4,6-triisopropylphényle

Figure 3.10.: 2-phenyl-2-tipyl acenapthenone and tipyl

e) *Analysis of the sample Dina5 or DN5*

The chromatogram of Dina5 indicates that the injected sample is a mixture of at least 14 compounds with at least 14 visible peaks (Figure 3.11).

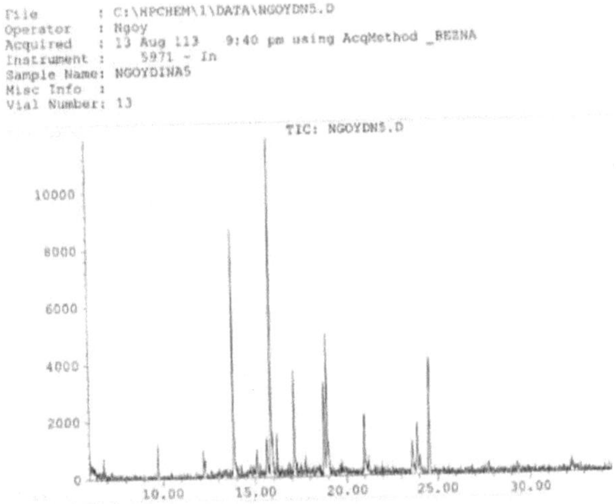

Figure 3.11.: Chromatogram of Dina5

GPC/MS data for the DN5 fraction compared to the literature and their mass spectra compared to their counterparts from the Wiley 275 L database indicate that :

- Peak 3 (13 821 min.) corresponds to **2E-nonadecene** (Beauchamp, 2013; NIST; SDBS; Cornu & Massot. , 1975) although the spectrum does not show the peak of the molecular ion. It was stopped before reaching the region of the peak of the molecular ion (M=266. 505).

- Peak 6 (15,864 min.) corresponds to **3E-eicosene** (Beauchamp, 2013; NIST; SDBS; Cornu & Massot 1975).

- Peak 10 (18.908 min) corresponds to **dioctadecyl phosphonate**.

The ^{1}H and ^{13}C-NMR data also confirm that this fraction is a mixture that contains a majority aliphatic hydrocarbon and at least one aromatic hydrocarbon by absorption in the region of the aromatic protons (**Figure 4.12. and 4.13 see Appendices**).

3.1.5.2. Analysis of DM samples from the methanolic extract in Ethyl acetate

a) Analyse the DM1 sample

According to DM1 GC/MS data, this fraction is a mixture of at least 10 products with ten peaks, seven of which are at 8,241min, 13,945min, 15,429min, 15,677min, 16,464min, 20,916 and 32,191min (Figure 3.12). This mixture was in the form of a grey powder from the methanolic extract washed in water and then recovered in ethyl acetate.

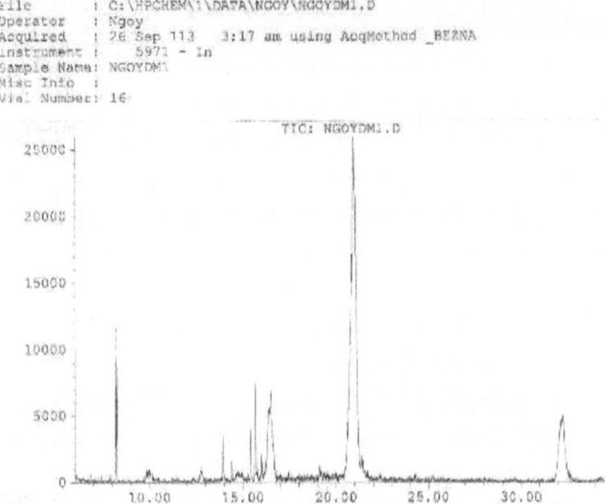

Figure 3.12.: Chromatogram of DM1

Only three compounds were identified by comparing their mass spectra with those in the database, namely :

- Peak 8 (15,677 min) is identified as **1,8-dimethoxy-3-methyl-3-anthracene-9,10-dione or 1,8-dimethoxy-3-methyl-anthraquinone.**

- The most intense peak 11 (20,916 min) corresponds to **triacontane**

- Peak 12 (32.191 min) corresponds to **hexatriacontane.**

The different fragmentations of some unidentified compounds are :
- Peak (13,945 min): 58, 71, 84, 97, 111, 133,141.

- Peak (15,429 min): 82, 101, 179, 207, 225, 267, 270, 282.

- Peak (16,464 min): 57, 71, 97, 134, 141, 164, 224, 241, 352, 394.

- Peak (18.242 min): 60, 61, 77, 78, 87, 94, 105, 122, 134, 150, 152

b) Analysis of the DM2 sample

The $^{1H\text{-}NMR}$ spectrum of the DM2 sample (Figure 4.12 in the Appendices) indicates that this compound is a mixture in which the majority product is an alkane. This is confirmed by the $^{13C\text{-}NMR}$ data (Figure 4.15 in the Appendices). All of these data suggest that this compound is *n-eicosane*.

Table 3.15. NMR spectrum of DM2

$^{1H\text{-}NMR}\delta_H(ppm)$:   0.80 (bs)   1.19(s)   0.80(bs)

$CH3 - _{CH2} - _{CH2} - _{CH2} - _{CH2}-(_{CH2})_{12}- _{CH2} - _{CH2} - _{CH2} - _{CH2} - _{CH2}- CH3$

$^{13C\text{-}NMR}\delta_C(ppm)$:   14.1   22.6   31.9   29.3   29.7   29.3   31.6   22.6 14.1

GIC/MS data for the DM2 fraction indicate that it contains at least 23 products (**Figure 3.13**). Comparison of the mass spectra of some of these products with data from the literature (**SDBS**) identified them. Thus :
- Peak 1 (11,616 min) corresponds to **2,6-di-tert-butylphenol**,

- Peak 2 (12,244 min) corresponds to **n-heptadecane** (Beauchamp, 2013, NIST, SDBS, Cornu *et al.*, 1975).

- Peak 4(13,822 min) corresponds to **α-octadecene**

- Peak 5 (14.302 min) has not been identified, its fragmentation by mass spectrometry is: 58, 71, 85, 95, 110, 124, 153, 165, 250.

- Peak 16 (18,919 min) corresponds to **1-heptadecene.**

- peak 20 (19,532min) could be identified with an isomer of **2-phenyl-2-tipyl acetophenone which** is the **2-phenyl-2-tipyl-2-acenaphthenone that has** also been identified in **Dina4** whose structure is shown in Figure 3.10 (Beauchamp ,2013).

Some fragments of 2-phenyl-2-tipyl-2-acenapthenone:

446: M+
431: M-15 (M-CH3)
403: M-43 (M-CH(CH3)$_2$
241: M-205 (M-tipylH+)

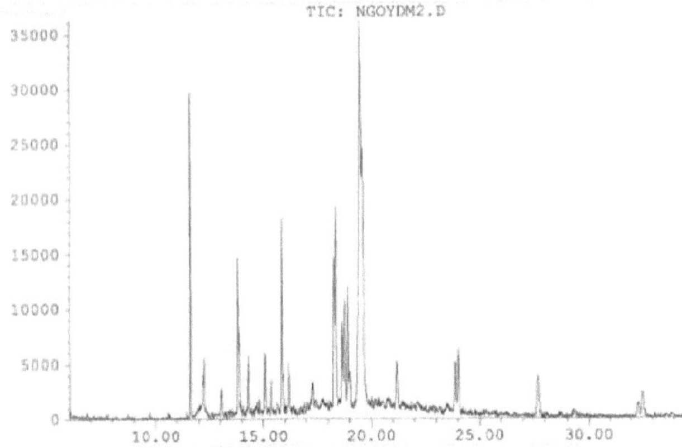

Figure 3.13.:   Chromatogram of the DM2 sample

c) **Analysis of the DM3 sample**

The chromatogram of DM3 isolated as a grey powder indicates that the injected sample is a mixture of at least 14 compounds with at least 14 peaks (Figure **3.14).**

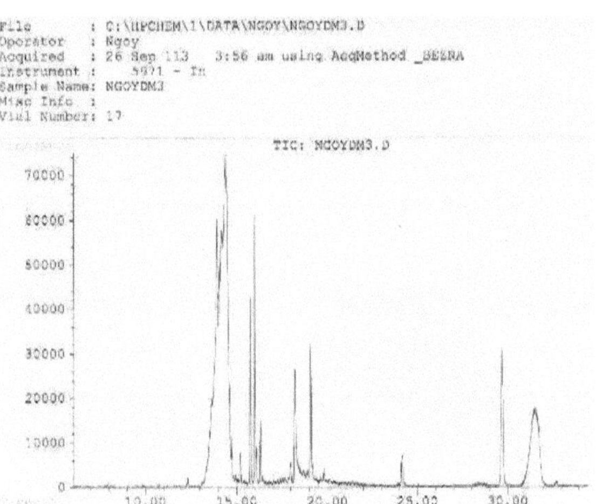

Figure 3.14. DM3 Sample Chromatogram

GC/MS analysis was performed on the DM3 sample and the mass spectra recorded for eight compounds were compared to those in the database for their identification. Thus :

- Peak 1(13.949min) corresponds to **tritetracontane,**
- Peak 2 (14.217 min) corresponds to **triacontane,**
- Peak 3 (14,395 min) corresponds to **tetradecane,**
- The 5(15.759min) peak corresponds to **benzene dibutyl-1,2-dicarboxylate** or **butylphthalate,**
- Peak 7 (16.007min) corresponds **to docosene-1**
- Peak 8 (16,322 min) corresponds to **methyl hexadecanoate**

- Peak 11 (19,097min) corresponds to **eicosanol-1**
- Peak 14 (29,672 min) is identified as **benzene bis (2-ethylhexyl) or bis (2-ethylhexyl) diterbutylphenol phthalate 1,2-dicarboxylate**.

3.1.5.3. Analysis of the Tsh-DO5/3/2 sample from the ethyl acetate extract

The chromatographic analyses (TLC and column) led to the obtaining of the product Tsh-Do5/3/2 in yellowish powder form, the TLC monitoring of which, using DCM as eluent and 20% sulphuric acid in ethanol as developer followed by heating at 110°C, showed a spot of persistent pink colour located at a value of Rf = 0.51.

1H-NMR of **Tsh-Do5/3/2**

The [1H-NMR] spectrum of M-TS-DO-5-3 (Figure 3.15) indicates the presence of singlet in the methyl group region, as well as peaks with an integration area corresponding to a proton at δ= 2.20 ppm, 3.43 ppm and 5.29 ppm. There is also a significant peak (one singlet) at δ=2.88ppm.

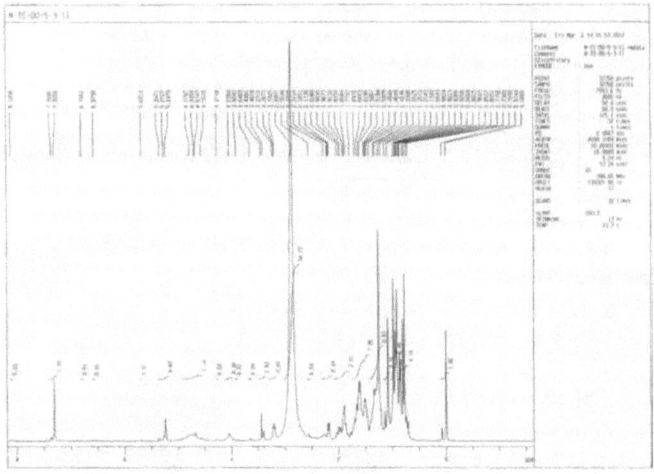

Figure 3.15. : 1H-NMR spectrum of Tsh-Do-5-3

The expansion of the region where methyl groups appear on the 1H-NMR spectrum (Figure 3.16) indicates the presence of two doublets at δ=0.86 ppm (J=6.5 Hz) and 0.94 ppm (J=6.5 Hz) each corresponding to a methyl group attached to a CH (-CH-CH3). There are also five singlets at δ=0.78, 0.81, 0.92, 0.98 and 1.08 ppm corresponding to five angular methyl groups. A weak singlet at δ=1.14 ppm (less than 3H), characteristic and a large singlet at δ=1.26 ppm are also observed. The peak at δ=1.14 ppm is generally characteristic of the 27-methyl group of an oleanene; the peak at δ=1.26 ppm reveals the presence of an aliphatic chain -$(CH_2)_n$- in M-TS-DO-5-3.

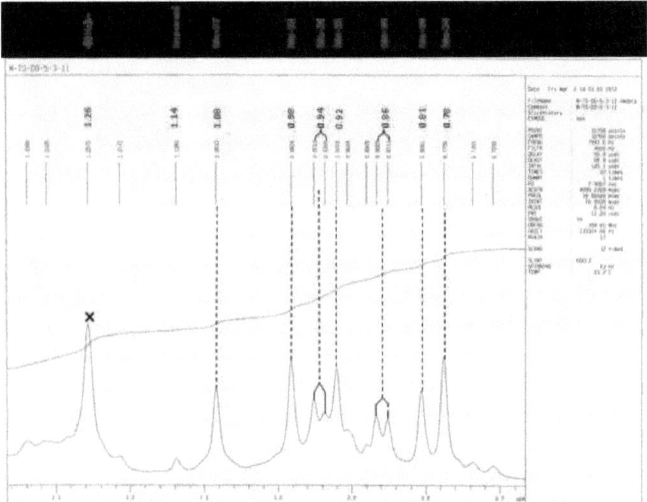

Figure 3.16.: Expansion of the [1H-NMR] Spectrum of Tsh-Do-5-3

The [13C-NMR] spectrum (Figure 3.17.) indicates in particular the presence of the following peaks, characteristic of triterpenes of - and β-amyrin groupsα: peaks at δ= 79.1, 122.3, 125.5, 138.1 and 181.0 ppm. The peak at δ= 79.1 ppm can be attributed to the C-3 carbon (CH-O-) of a triterpenoid; peaks 122.5 and 138.1 ppm are characteristic of the double bond Δ12 of ursane-type triterpenoids (α-amyrin) (Sang *et al.*, 2002).

The peak at δ=122.3 ppm is characteristic of the C-12 carbon of the 12 double bond of Δoleanane-type triterpenoids (β-amyrin) although the characteristic C-13 carbon peak of this double bond is not observed. Finally, the peak at δ=181 ppm corresponds to a free carboxylic group (-COOH).

These spectroscopic data indicate that M-TS-DO-5-3 is a mixture of the ursane and oleanane-type triterpenes. They also suggest the presence of a fatty acid or its ester, i.e. any alkane.

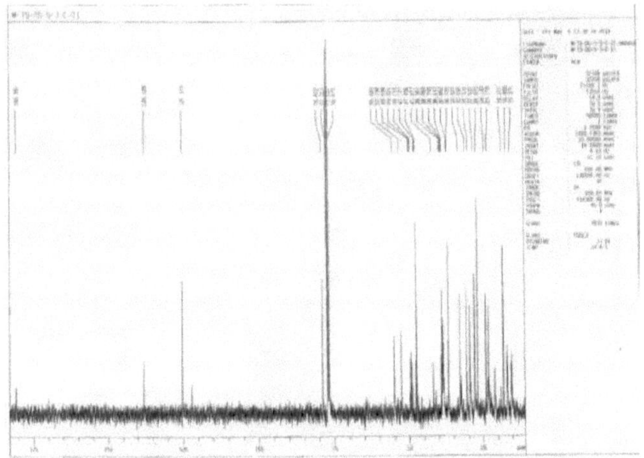

Figure 3.17.: 13C-NMR spectrum of Tsh-Do-5-3

Data for [1H], [13C], 2D-NMR of the majority product in Tsh-DO-5-3 are grouped in Table 3.16 below.

Table 3.16. : ^1H and ^{13}C NMR data

C	^1H (δ)	Multiplicity	^{13}C (δ)	DEPT	C	^1H (δ)	Multiplicity	^{13}C (δ)	DEPT
1			38.7	CH2	16			24.3	CH2
2			28.1	CH2	17			43.8	C
3	3.43	(1H, *dd*, *J*= 10.0 Hz, 4.5 Hz)	79.1	CH	18	2.20	(1H, *d* , *J=11.7 Hz,*)	52.7	CH
4			39.0	C	19			39.2	CH
5	0.72	(1H, *m*)	55.3	CH	20			38.9	CH
6			18.4	CH2	21			30.7	CH2
7			33.1	CH2	22			36.9	CH2
8			39.4	C	23	0.98	3H, *s*	28.1	CH3
9			47.6	CH	24	0.78	3H, *s*	15.7	CH3
10			36.7	C	25	0.92	3H, *s*	15.5	CH3
11			23.3	CH2	26	0.81	3H, *s*	13.0	CH3
12	5.29	(1H, *t*, *J=3.6 Hz*)	125.5	CH=	27	1.08	3H, *s*	23.6	CH3
13			138.1	C=	28	-----		181.0	O-C=O
14			41.3	C	29	0.86	3H, *d* : J=6.5 Hz	13.1	CH3
15			28.1	CH2	30	0.94	3H, *d* : J=6.5 Hz	21.2	CH3

1. Majority Compound

In the high field region, the [1H-NMR] spectrum (Table 3.16.) indicates the presence of seven methyl groups, including two secondary methyls and five singlets suggesting a skeleton of an ursan triterpene. In the weak field region, the signals at δ=3.43 and 5.29 ppm are logically attributed to protons attached to carbons C-3 (bearing an OH) and C-12 respectively.

Data from [13C-NMR] and DEPT [13C-NMR] (Table 3.16) indicate the presence of a 12 double bond at $\Delta\delta$=125.5 (=CH) and 138.1 (-C=) ppm, confirming that this compound belongs to the ursane group (Sang *et al.*, 2002).

The DEPT [13C-NMR] (Figures 3.18 and 3.19) also reveals that this molecule has 7 CH3, 9 $_{CH2}$, 7CH, and 7 C (quaternary carbons). The signal at δ= 79.1 ppm corresponds to carbon C-3 (carrying an HO) and the peak at δ=181 ppm indicates the presence of a -COOH group attributed to carbon C-28. All these data suggest that the majority compound is an ursane-type triterpene acid.

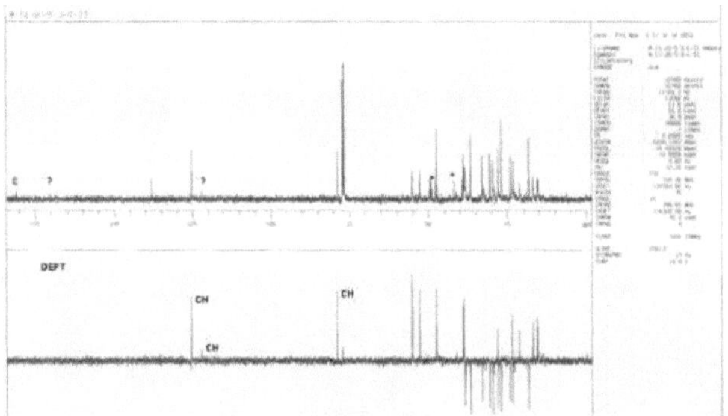

Figure 3.18. Tsh-Do-5-3 majority compound NMR-DET-NMR compared to 13C-NMR

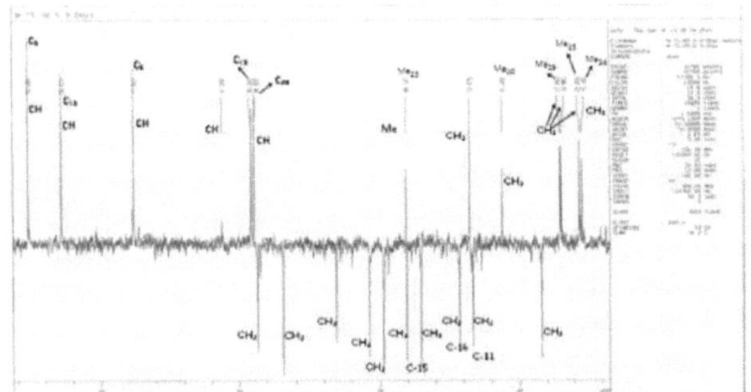

Figure 3. 19 : Tsh-Do-5-3 majority compound NMR-DEPT

The allocation of the other carbons was based on [1H] and [13C-NMR] correlation data (Table 3.17 and Figures 3.20 and 3.21). The COSY spectra indicate correlations between protons separated by two or three bonds. Thus, the COSY spectrum reveals correlation peaks between H-3 and H-2 (δ=1.60ppm), H-12 and H-11 (1.90 ppm), H-11 and H-9 (1.50 ppm), H-18 and H-19 (1.38 ppm), H-19 and H-20 (1.5 ppm).

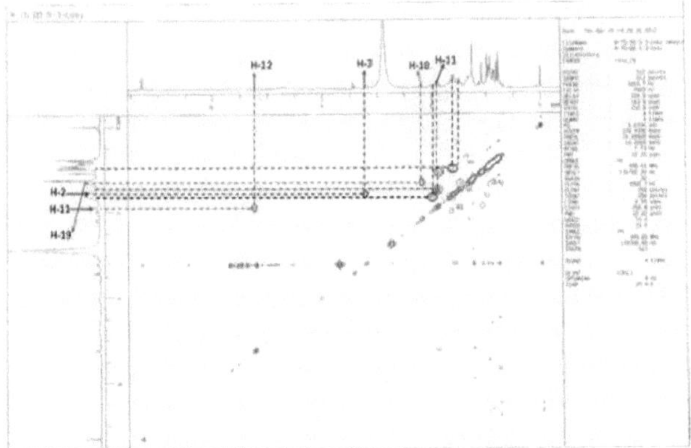

Figure 3.20.: COSY-NMR of the majority Tsh-Do-5-3 compound

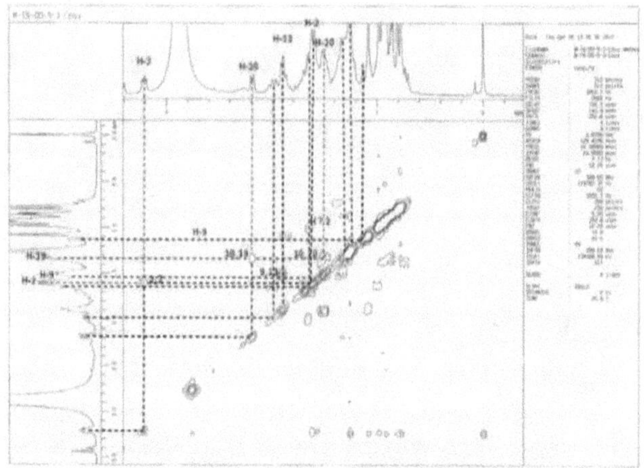

Figure 3.21. : COSY-NMR expansion of the majority Tsh-Do-5-3 compound

Table 3.17: [1H] and 13C-NMR Correlations

C	δ_H	δ_C	COSY	HMQC	HMBC	C	δ_H	δ_C	COSY	HMQC	HMBC
1						16	1.80 (H, *m*)	24.3		C-16	
2	1.60 (1H, *m*)	28.1	H-3	C-2	C-3	17	------				
3	3.43 (1H, *dd*, *J*= 10.0 Hz,	79.1	H-2	C-3	C-23, C-24	18	2.20 (1H, *d*, *J=11.7* Hz,)	52.7	H-19	C-18	C-12, C-13, C-14, C-17, C-18, C-20
4	------	39.0				19	1.38 (1H, *m*)		H-18, H-20	C-19	
5	0.72 (1H, *m*)	55.3	H-5	C-5		20	1.50 (1H, *m*)		H-19, H-20		
6						21	1.59 (H, *m*)	30.7		C-21	
7						22	1.60 (H, *m*), 1.70 (H, *m*)	36.9		C-22	
8						23	0.98 (3H: s)	28.1			C-3, C-4, C-5, C-24
9	1.50 (1H, *m*)	47.6	H-11	C-9	C-8, C-10, C-11, C-25, C-26	24	0.78 (3H: s)	15.7			C-3, C-4, C-5, C-23
10						25	0.81 (3H: s)	15.5			C-1, C-5, C-9,
11	1.90 (1H, *m*)	23.3	H-9	C-11	C-9, C-12, C-13	26	0.92 (3H: s)	13.0			C-7, C-9, C-14
12	5.29 (1H, *t*, *J=3.6* Hz)	125.5	H-11	C-12		27	1.08 (3H: s)	23.6			C-8, C-13, C-14, C-15
13						28	------				
14						29	0.86 (3H, *d* : *J* =6.8 Hz)	13.1			C-18, C-20
15						30	0.94 (3H, *d* : *J* =6.8 Hz)	21.2			C-19, C-21

HMQC spectra indicate a correlation between a proton and the carbon to which it is fixed. Thus, HMQC data for the majority product indicate correlation peaks between protons and the following carbons: H-3 (3.43 ppm) and C-3 (79.1 ppm), H-5 (0.72 ppm) and C-5 (55.3 ppm), H-6 (1.58 ppm) and C-6 (18.4 ppm), H-9 (1.50 ppm) and C-9 (47.6 ppm), H-12 (5.29 ppm) and C-12 (125.5 ppm), H-18 (2.20 ppm) and C-18 (52.7 ppm). Other HMQC correlations are shown in Table 3.17 and Figures 3.22. and 3.23.

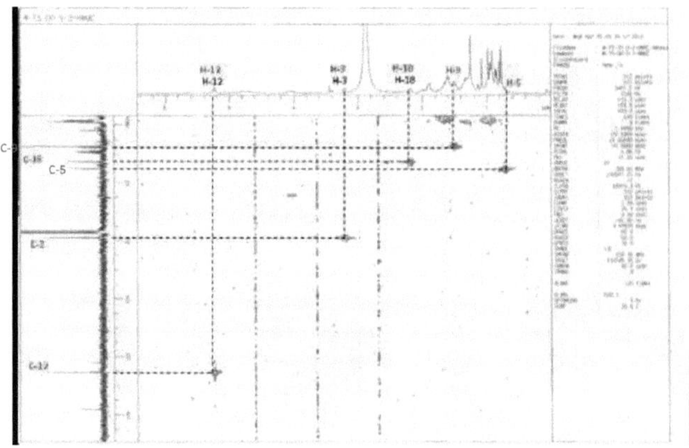

Figure 3.22. HMQC-NMR of the majority Tsh-Do-5-3 compound

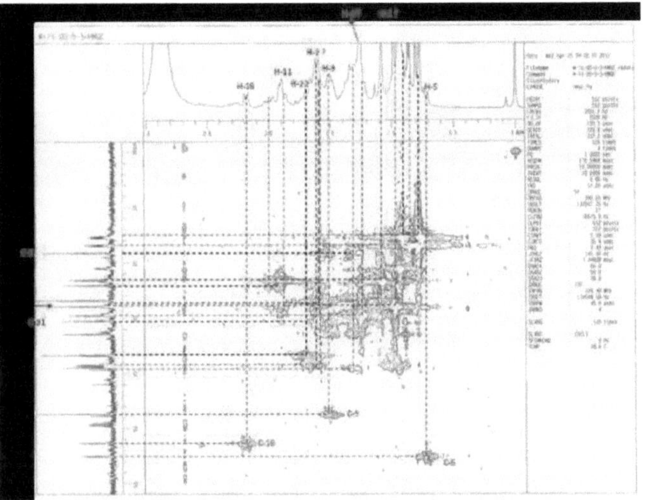

Figure 3.23.: HMQC-NMR expansion of the majority Tsh-Do-5-3 compound

HMBC spectra indicate correlation peaks between protons and carbons separated by two or three bonds. The HMBC data (Table 3.17 and Figures 3.24-3.26) show the following correlations:

- H-3 (3.43 ppm) and C-23 (28.1 ppm) and C-24 (15.7 ppm)
- H-18 (2.20 ppm) and C-12 (125.2 ppm), C-13 (138.1 ppm), C-14 (41.3 ppm), C-16 (24.3 ppm), C-17 (43.8 ppm), H-20 (38.9 ppm) and C-29 (13.1 ppm)
- H-11 (1.90 ppm) and C-9 (43.6 ppm), C-12 and C-13.
- H-9 (1.50 ppm) and C-8 (39.4 ppm), C-10 (36.7 ppm), C-25 (15.5 ppm) and C-26 (13.0 ppm)
- H-5 (0.72 ppm) and C-3 (79.1 ppm), C-24 and C-25 (15.5 ppm)
- H-23 (0.98 ppm) and C-3, C-4 (39.0 ppm), C-5 (55.3 ppm), C-24
- H-24 (0.78 ppm) and C-3, C-4, C-5, C-23
- H-25 (0.0.92 ppm) and C-1, C-5, C-9

- H-26 (0.81ppm) and C-7 (33.1 ppm), C-8, C-14,
- H-27 (1.08ppm) and C-5, C-7 (33.1 ppm), C-8, C-9, C-14,
- H-25 (0.81ppm) and C-5, C-7 (33.1 ppm), C-8, C-9, C-14,
- H-29 (0.86ppm) and C-18, C-20
- H-30 (0.94ppm) and C-19, C-21

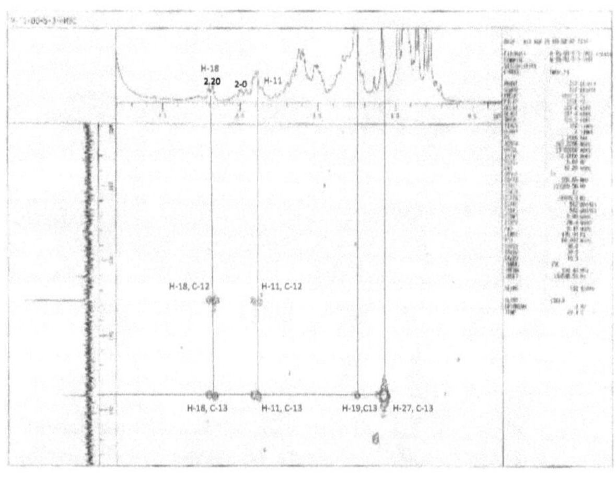

Figure 3.24. HMBC-NMR of the majority Tsh-Do-5-3 compound

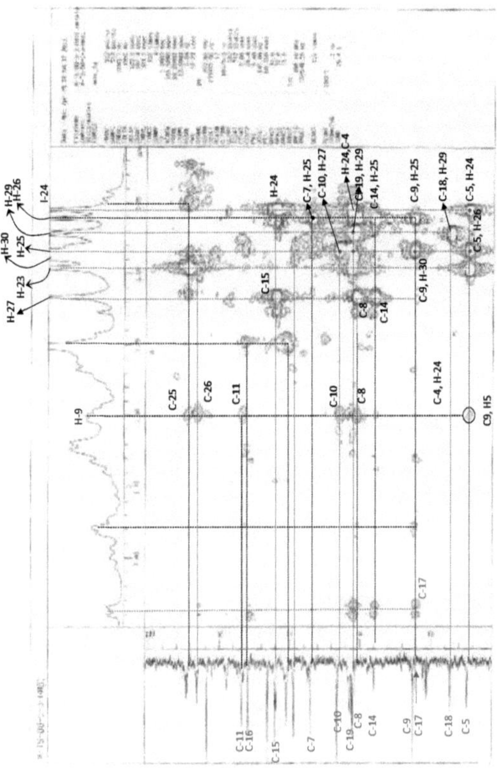

Figure 3.25. HMBC-NMR of the majority Tsh-Do-5-3 compound (continued 1)

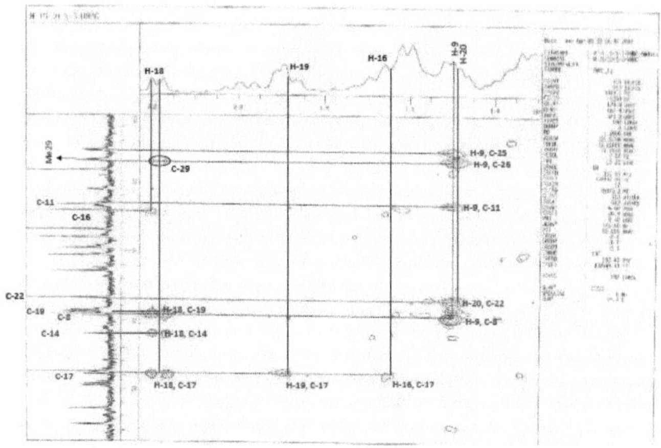

Figure 3.26. HMBC-NMR of the majority Tsh-Do-5-3 compound (continued2)
These correlations clearly identify this compound with ursolic acid.

NMR is the best method for distinguishing the types of triterpenes. In [1H-NMR], for oleanane-type triterpenes all methyl groups are tertiary and they absorb in the form of singlets. The ursane-type triterpenes have, in addition to the tertiary methyls, two secondary methyls which absorb in the form of doublets.

In [13C-NMR], the distinction between these two types of triterpenes is in the values of the C-12 and C-13 carbons of the double bond. Values between 124 and 139 ppm are characteristic of the ursane-type, while those between 122 and 144 ppm indicate the presence of an oleanane-type structure.

All NMR data for this compound identify it as ursolic acid. This is also confirmed by mass spectrometry data. Indeed, the mass spectrum (figures 3.27. and 3.28.) reveals the peak of the molecular ion at m/z = 456 which corresponds to the molecular formula C30H48O3. The base peak at m/z = 248 and a less intense peak at m/z 207 are consistent with the RDA (Retro Diels-Alder) fragmentation of the C-ring of a triterpenoid of the -amyrin (urethane) or β-amyrin (oleanene) type (αSeebacher et *al.*, 2003).

The remaining peaks are formed following the fragmentation process shown in Figure 3.29. On the basis of all these data, compared to those in the literature, the majority compound has therefore been identified as **ursolic acid** (Tshilanda *et al.*, 2015b).

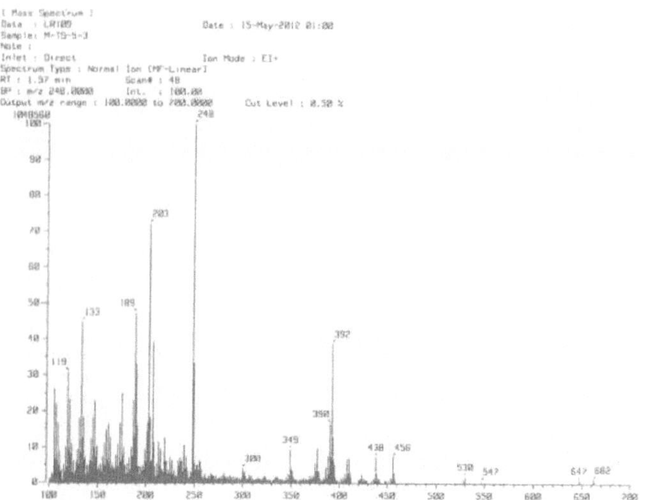

Figure 3.27. : Mass spectrum of the majority compound of Tsh-Do-5-3

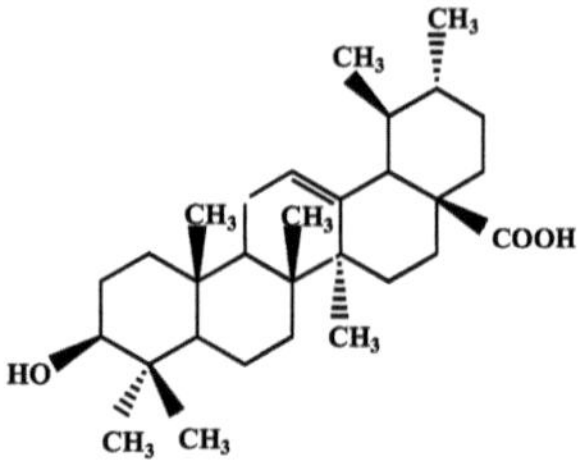

Figure 3.28. : Ursolic acid

The fragmentation of ursolic acid in mass spectrometry is shown in Figure 3.42. below:

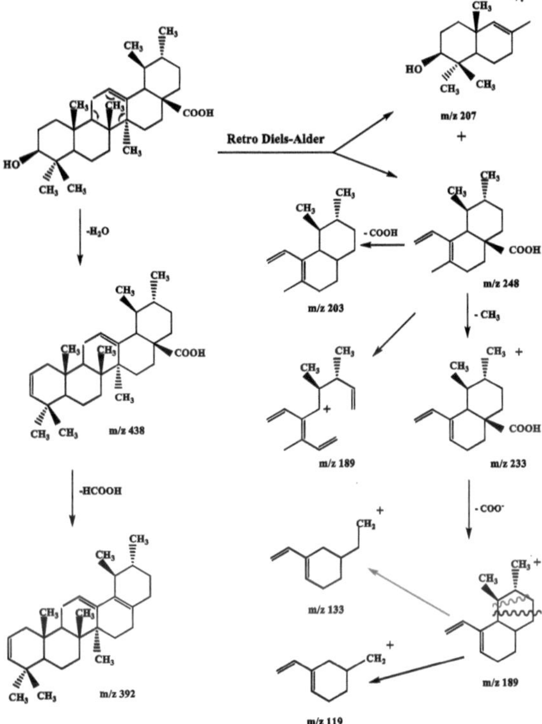

Figure 3.29. Fragmentation of the majority compound in mass spectrometry

2. Minority compound

Although all of these spectroscopic data indicate that M-TS-DO-5-3 consists essentially of ursolic acid, they also indicate that this compound is contaminated with a trace amount of an oleanane-type triterpene and a fatty acid ester. Indeed, the NMR indicates, among other things, a small peak at $\delta=122.3$ ppm characteristic of the carbon 12 of a double bond $\Delta 12$ of a triterpene of the oleanane-type in [13C-NMR] and a low intensity peak at δ-1.14 ppm characteristic of the methyl group at position 27 of the same oleanene type in [1H-NMR] (Seebacher et *al.*, 2003).

The presence of a large peak (a singlet) at $\delta=1.26$ ppm, characteristic of a set of methylene groups - $(CH_2\text{-})n$- in the chain of an alkane, a long chain alcohol or a fatty acid or fatty acid ester in [1H-NMR] is also observed. The mass spectrum of M-TS-DO-5-3 shows a low intensity peak at m/z = 662 ($M^{.+}$), followed by a peak at m/z = 647 (M-CH3). Table 3.18. **shows** some of the peaks produced by the fragmentation of M=662.

Table 3.18. : Some Peaks in the Mass Spectrum of the Minority Product

Peak (m/z)	Intensity (%)	Fragment
662	1	$M.^{+}$
647	0.7	M-15 (CH3)
423	2	M-239 (CH3-$(CH2)_{14\text{-CO-}}$)
408	7	M-254 (CH3-$(CH2)_{14\text{-CO2}}$, -H+)

These data suggest that this minority compound can be identified with the **palmitate of a triterpenoid** whose structure has yet to be determined.

3.1.5.4. Analysis of the DOR4 sample from the extract at DCM

The study of the compound DOR4 was exclusively done on the basis of [1H] and [13C]-NMR data. The [1H-NMR] spectrum (Table 3.19. and Figures 3.30 and 3.31) shows a high number of peaks in the strong field region ($\delta = 0.69$-1.41 ppm). Most of these signals correspond to the peaks (singlet) characteristic of the tertiary methyl groups of the triterpenoids.

In the weak field region there are three peaks centered at : $\delta = 2.75$ ppm (1H dd), $\delta= 3.15$ ppm (1H dd) and 5.22 ppm (1 H, m). There is also the presence of a significant peak at $\delta= 1.18$ ppm (singlet) characteristic of methylene groups [$-(CH_2)n-$] of an aliphatic chain.

The [13C-NMR] spectrum (Figures 3.32 and 3.33) of Dor4 indicates, in addition to other peaks, the presence of peaks at $\delta= 79.1$ ppm (-CH-O), 122.6 ppm, 143.3 ppm, and 181.2 ppm. The olefinic carbon signals at $\delta= 122.6$ ppm (C-12) and 143.3 ppm (C-13) are characteristic of a 12 double bond of Δan oleanene (Seebacher et *al.*, 2003), and the signal at $\delta= 181.2$ ppm reveals the presence of a COOH group in DOR 4. These signals, combined with the methyl group peaks observed at [1H-NMR] suggest that DOR 4 contains an oleanane-type triterpene.

Table 3.19. : [1H] and [13C] NMR data

C	^{1H} (δ)	Multiplicity	^{13C} (δ)	C	^{1H} (δ)	Multiplicity	^{13C} (δ)
1			38.5	16			22.7
2		1.55(*m*)	23.2	17			46.5
3	3.15	(1H, *dd*, *J*= 10.0, 4.5 Hz)	79.1	18	2.75	(1H, *dd*, *J*= 13.2, 4.2Hz,)	42.0
4			38.8	19			43.7
5	0.65	(1H, *m*)	55.3	20			30.7
6			18.3	21			33.8
7			32.5	22			31.9
8			39.3	23	0.92	3H, *s*	28.2
9			43.9	24	0.71	3H, *s*	15.6
10			33.1	25	0.86	3H, *s*	15.3
11			23.4	26	0.71	3H, *s*	13.1
12	5.22	(1H, *m*)	122.3	27	1.07	3H, *s*	25.9
13			143.3	28	-----		181.2
14			41.3	29	0.83	3H, *s*	33.1
15			23.7	30	0.84	3H, *s*	23.6

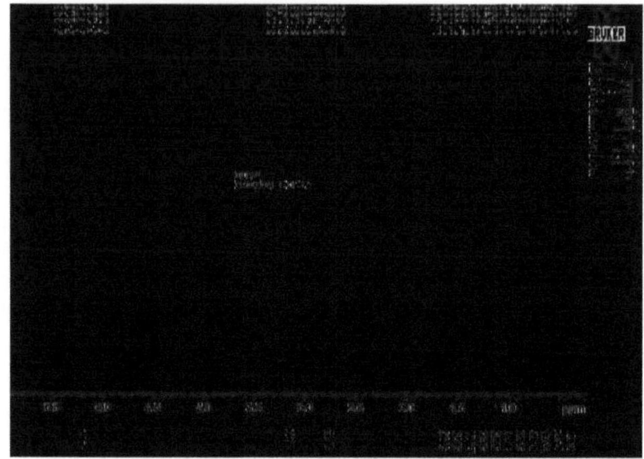

Figure 3.30. : 1H-NMR spectrum of SOR 4

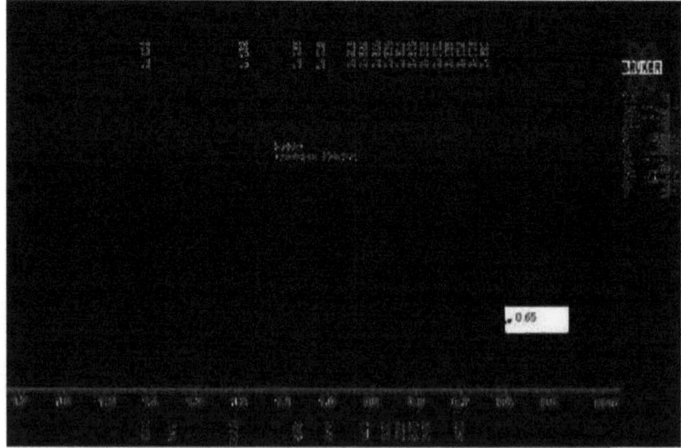

Figure 3.31. Expansion of the methyl group region of the 1H-NMR spectrum of DOR4

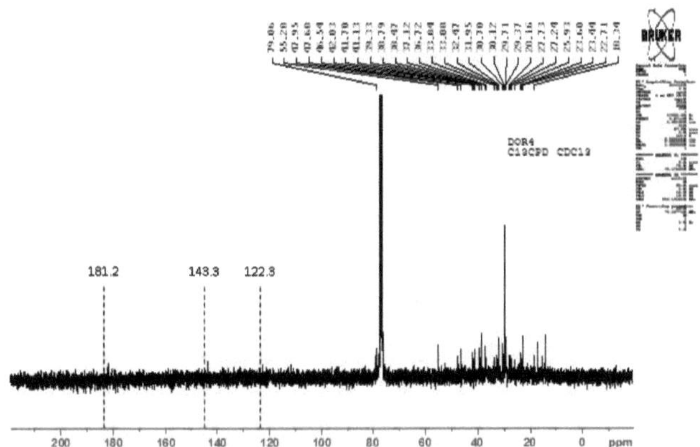

Figure 3.32. : $^{13C-NMR}$ spectrum of SOR 4

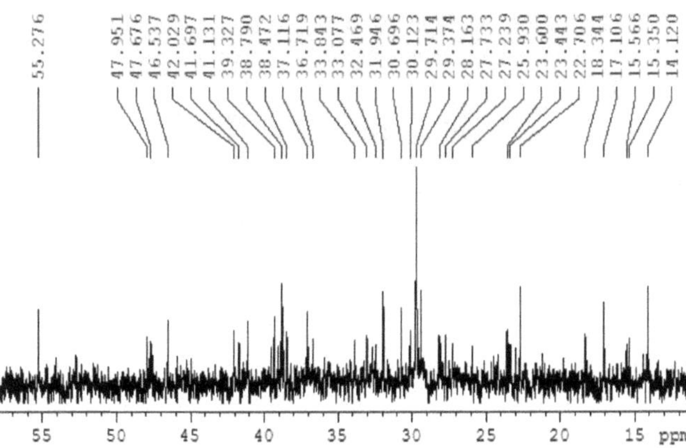

Figure 3.33. Expansion of the 0-55 ppm region of the $^{13C-NMR}$ spectrum of SOR 4

Examination of the 2D-NMR correlation data COSY, HMQC and HMBC confirm that DOR4 is an oleanane-type triterpene. Indeed, the $^{1H-1H-COSY}$ spectrum (**Figure 3.34.**) indicates that the absorbing proton at δ= 2.75 ppm (H-18) is coupled to the proton appearing at δ= 1.10 ppm (H-19); and that the peak δ=3.15 ppm (H-3) is coupled to the absorbing proton at δ= 1.58 ppm (H-2).

It also shows that there is a correlation between the signal at δ= 5.22 ppm (H-12) and the peak at δ= 1.85 ppm (H-11). The signal at δ= 1.85 ppm (H-11) correlates with the absorbing proton at δ= 1.45 ppm (H-9) of an oleanene-type triterpene.

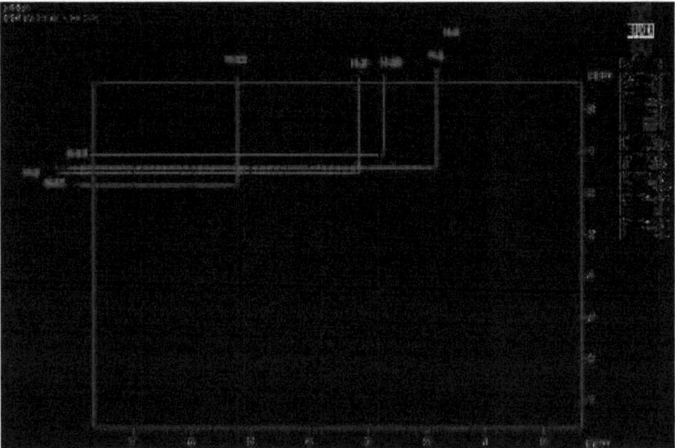

Figure 3.34. 2D-COSY correlations of DOR 4

The HMQC spectra (Figures 3.35, 3.36 and 3.37.) report correlations between protons and the carbons to which they are attached. Thus, the signal centred at 3.15 ppm (H-3) is attached to the absorbent carbon at δ= 79.1 ppm (C-3); the signal centred at =δ 5.22 ppm (H-12) is coupled to the carbon at δ= 122.3 ppm (C-12). There is also a correlation between the proton at δ= 1.85ppm (H-11) and the carbon at δ= 23.3 ppm (C-11).

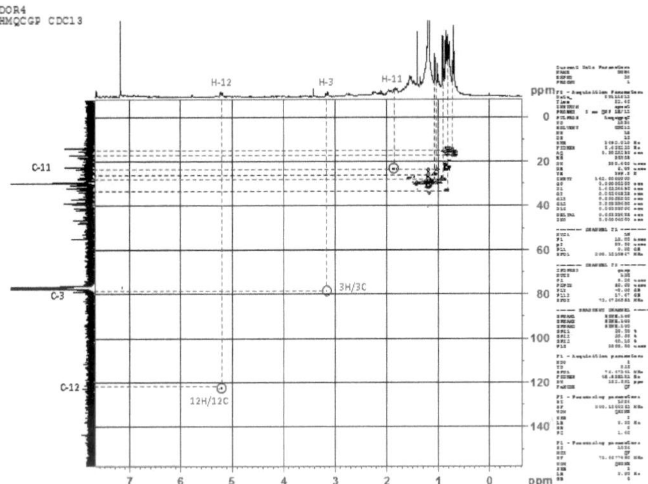

Figure 3.35. 2D-HMQC Correlations of SOR 4

There are also correlations between the protons of the methyl groups and the carbons to which they are attached.

Thus, the signal at δ= 0.71ppm shows couplings with carbon at δ= 15.6 ppm (methyl 24) and 13.1 ppm (methyl 25). The peak at δ= 0.83 ppm is coupled to carbon at δ= 33.1 ppm (methyl 29) and the proton signal at = δ0.84 ppm correlates with carbon at δ= 23.6 ppm (methyl 30). The proton signal absorbing at δ= 0.92 ppm is coupled to the carbon at δ= 28.2 ppm (methyl 23). Finally the peak at δ= 1.07 ppm correlates with carbon at δ= 25.9 ppm (methyl 27).

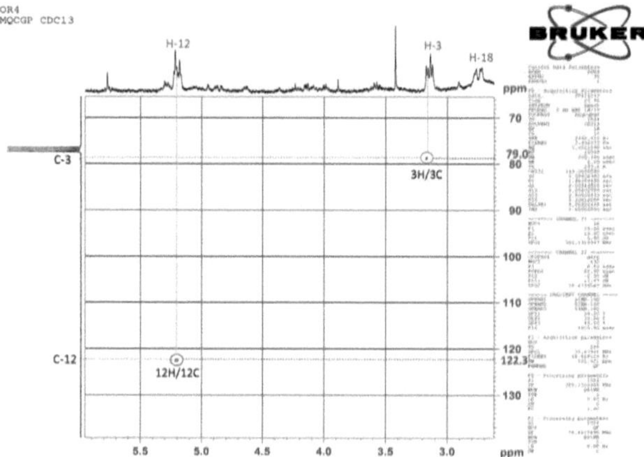

Figure 3. 36.: Expansion 1 2D-HMQC Correlations of SOR 4

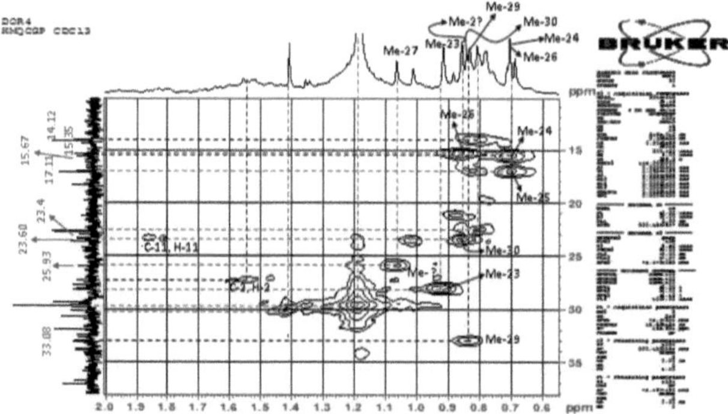

Figure 3.37. : Expansion 2 2D-HMQC Correlations of DOR 4

Recall that the HMBC spectra indicate couplings between protons and carbons separated by two or three bonds. The HMBC spectrum of DOR 4 (Table 3.19. and Figure 3.38.) corroborates the other spectral data already analysed. Thus, protons at δ= :

- 0.71 ppm: coupled with carbons C-3, C-5, and C23 (methyl 24)
- 0.71 ppm: coupled to C-7, C-8, and C14 (methyl 25) carbons
- 0.83 ppm: coupled with carbons C-19, C-20, and 21 (methyl 29)
- 0.84 ppm: coupled with carbons C-19, C-20, and C-21 (methyl 30)
- 0.86 ppm: coupled to C-5, C-9, and C-1 (methyl 26) carbons
- 0.92 ppm: coupled with carbons C-3, C-5, C-4 and C-24 (methyl 23)
- 1.07 ppm: Clearly coupled to the C-8, C-13, C-14, and C-15 carbons (methyl 27).

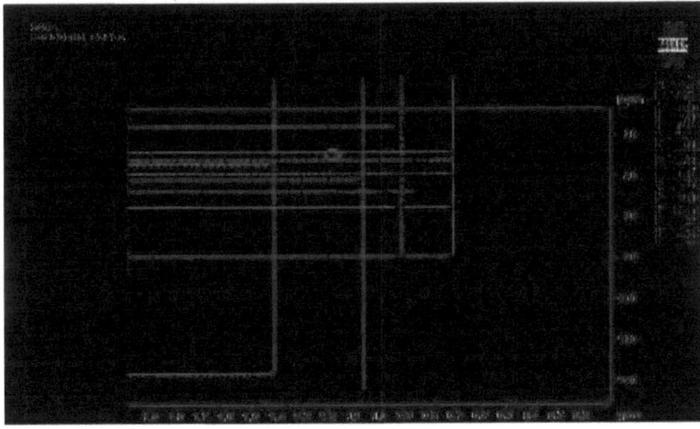

Figure 3.38. 2D-HMBC Correlations of SOR 4

These data, when compared to data in the literature, confirm that DOR 4 is predominantly composed of oleanolic acid (Ragasa & Lim, 2005).
They also indicate the presence of other minority compounds.

Indeed, the [1H-NMR] spectrum (**figure 3.30.**) shows other peaks characteristic of methyl groups (0.69, 0.72, 0.78, 0.79, 1.01 and 1.41 ppm) and a large peak (a singlet) at δ= 1.18 ppm which is coupled only to the absorbent carbons at δ= 29.7 ppm in HMQC and HMBC (**figures 3.37 and 3.38).** It is characteristic of the -$(CH_2)n$- part of a chain of a saturated aliphatic hydrocarbon. The presence of an alkane chain is also confirmed by the peak
at δ= 14.1 ppm in [13C-NMR] characteristic of the CH3- terminal group of such chains (***CH3*-**(CH_2)n-) (Charles *et al.,*1993).

In HMQC a coupling between carbon at δ=14.1 ppm (CH3-terminal) and protons at δ= 0.82 ppm, the absorption region of this kind of methyl in [1H-NMR]. There is also a correlation between the signal at δ= 1.41 ppm and the olefinic carbon signal at δ= 136.3 ppm suggesting that one of the minority products of DOR4 could be a triterpene of ursane type (Δ12 between 125 and 138 ppm).

As there is insufficient data on these minority products, their identification has not been possible.

Table 3.20.: [1H-NMR] and 13C-NMR Correlations of DOR4

C	δ_H	δ_C	COSY	HMQC	HMBC	C	δ_H	δ_C	COSY	HMQC	HMBC
1						16		22.7			
2	1.55	23.2	H-3	C-2	C-3	17	------	46.5			
3	3.15 (1H)	79.1	H-2	C-3	C-23, C-24	18	2.75 (1H)	41.1	C-19	C-18	
4	------	38.8				19	1.15 (1H)			C-19	
5	0.65 (1H)	55.3	H-5	C-5		20					
6						21					
7						22					
8						23	0.92 (3H: s)	28.2		C-23	C-3, C-4, C-5, C-24
9	1.50 (1H)	43.9	H-11	C-9		24	0.71 (3H: s)	15.6		C-24	C-3, C-5, C-23
10						25	0.86 (3H: s)	17.1		C-25	C-7, C-9, C-14 5, C-9,
11	1.85 (1H)	23.3	H-9	C-11		26	0.71 (3H: s)	15.3		C-26	C-1, C-5, C-9
12	5.22 (1H)	122.3	H-11	C-12		27	1.07 (3H: s)	25.9		C-27	C-8, C-13, C-14, C-15
13						28	------				
14						29	0.83 (3H, s)	33.1			C-18, C-20
15						30	0.84 (3H, s)	23.6			C-19, C-21

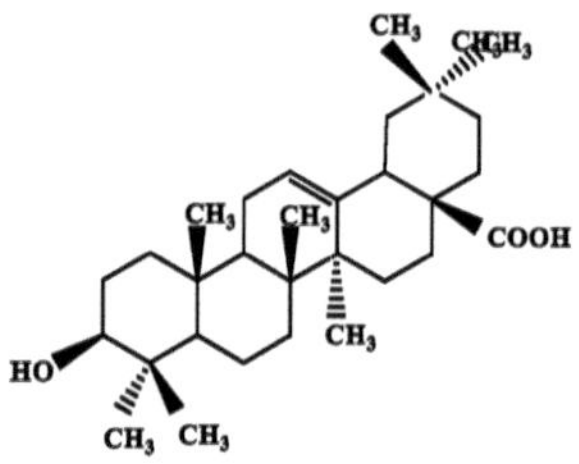

Figure 3.39. : Oleanolic acid

3.1.5.5. Analysis of the Dor2 sample from the extract at DCM

The DOR 2 sample was mainly analysed on the basis of [1H-NMR], [13C-NMR], DEPT and 2D-NMR correlations (COSY, HMQC and HMBC). It appears from this analysis that DOR2, which appears pure, is in fact a mixture of two hydrocarbons: an alkene and an alkane.

- **1H-NMR**

The [1H-NMR] spectrum of DOR 2 (Table 3.21. and Figures 3.40.and 3.41) shows two distinct regions: the olefinic proton region (5.02 - 5.08 ppm) and the aliphatic proton region (0.8 - 2.0 ppm).

In the region of the olefinic carbons, there is a signal (byte) corresponding to 3 olefinic protons (in agreement with the integration).

In the region of the aliphatic carbons, there are 4 peaks corresponding to 4 allylic methyls at δ: 1.53 ppm (*s*: 3 methyls) and 1.61 (*s*: 1 methyl). There are also two bytes centred at $\delta 1.94$ ppm and $1\delta.98$ ppm. Their integration indicates the presence of 10 protons (i.e. 5_{CH2}). In addition to these peaks, we also note the presence of a singlet at δ: 1.18 ppm corresponding to (9 protons) and a byte at δ: 0.86 ppm.

Table 3.21.: ^{1H} data - SOR 2 NMR

104

Woodpecker	δ (ppm)	Multiplicity
1	0.86	m
2	1.18	s
3	1.53	s
4	1.61	s
5	1.91 – 1.93	m
6	1.94 – 2.00	m
7	5.02 – 5.07	m

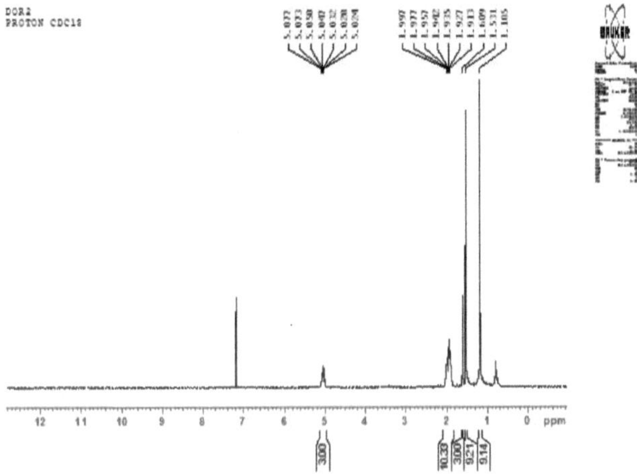

Figure 3.40. : ^{1}H-NMR spectrum of SOR 2

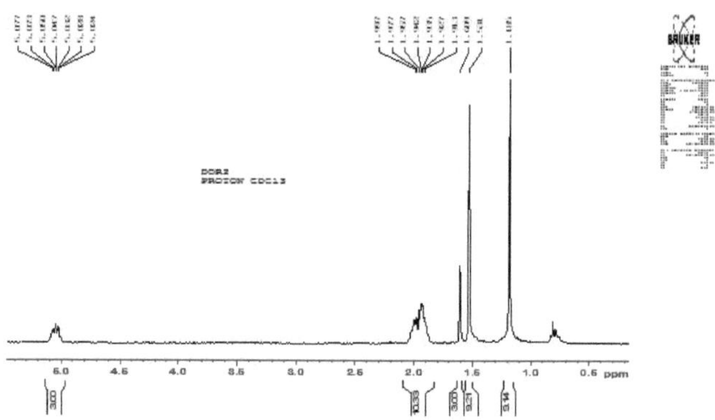

Figure 3.41. : Expansion of the [1H-NMR] spectrum of SOR 2

- [13C] - NMR

The [13C-NMR] spectrum of DOR 2 (Table 3.22. and Figure 3.42) also shows two distinct regions with 20 peaks: the olefinic carbon region (124.32 - 135.13 ppm) and the aliphatic carbon region (14.12 - 39.78 ppm).

In the olefinic carbon region, 6 carbons are observed, suggesting the presence of 3 carbon-carbon double bonds. The aliphatic region has 14 carbons.

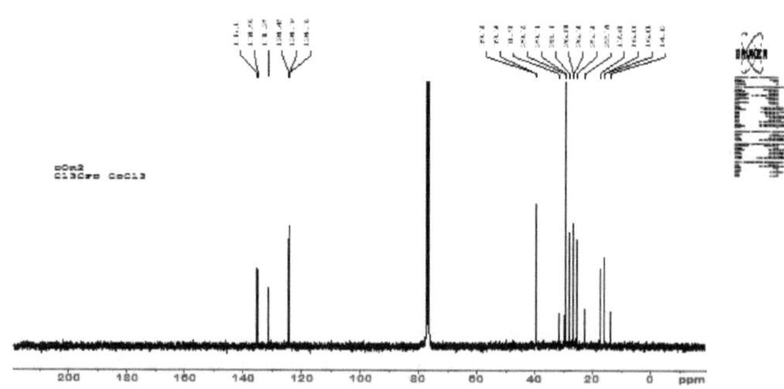

Figure 3.42. : [13C-NMR] spectrum of SOR 2

Table 3.22. : [13C-NMR] data from SOR 2

Woodpecker	Group	δ(ppm)	Woodpecker	Group	δ(ppm)
1	CH3	14.12	11	CH2	29.72
2	CH3	16.02	12	CH2	31.95
3	CH3	16.06	13	CH2	39.76
4	CH3	17.69	14	CH2	39.78
5	CH2	22.71	15	CH	124.32
6	CH3	25.70	16	CH	124.34
7	CH2	26.70	17	CH	124.45
8	CH2	26.81	18	C	131.25
9	CH2	28.30	19	C	134.92
10	CH2	29.38	20	C	135.13

- **[13C-NMR] DEPT**

The [13C-NMR] DEPT spectra (Table 3.23 and Figure 3.43) indicate that there are 3 methine groups (CH=) in the olefinic carbon region, suggesting the presence of 3 trisubstituted carbon-carbon double bonds. The peaks at δ: 131.25, 134.92 and 135.13 ppm are not observed on the [13C-NMR] DEPT spectra: they therefore correspond to quartenary olefinic C of the double bond (C = C - H).

In the region of the aliphatic carbons, there are 4 peaks corresponding to methyl groups (CH3) and 6 peaks corresponding to methylene groups (CH2).

Table 3.23.: SOR 2: 13C-NMR-DEPT

Woodpecker	δ(ppm)	DEPT	Woodpecker	δ(ppm)	DEPT
1	16.02	CH3	9	39.76	CH2
2	16.06	CH3	10	39.78	CH2
3	17.69	CH3	11	124.32	CH=
4	25.70	CH3	12	124.34	CH=
5	26.70	CH2	13	124.45	CH=
6	26.81	CH2	14	131.25	C=
7	28.30	CH2	15	134.92	C=
8	29.72	CH2	16	135.13	C=

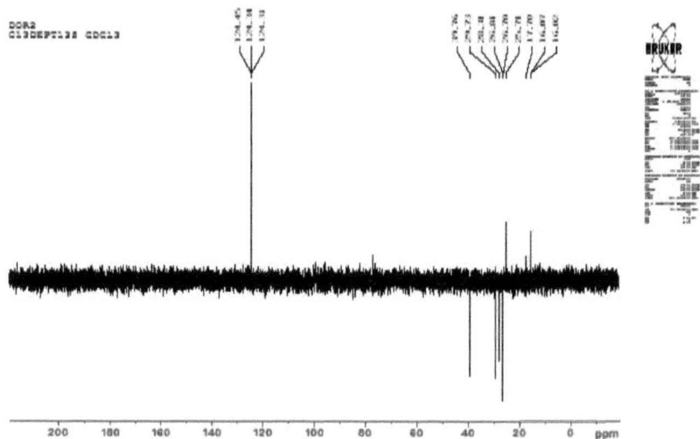

Figure 3.43.: DEPT $^{13C\text{-NMR}}$ spectrum of SOR 2

- **2D-NMR**

1) COSY Correlations

The COSY NMR spectrum (Figure 3.44) shows the correlations reported in Table 3.24. For clarity, the $^{1H\text{-NMR}}$ signals are classified into nine groups: A_1 (=5.03 ppm), A_2 (5.05 ppm), A_3 (5.07 ppm), B (1.99-2.09 ppm), C (1.88-1.98 ppm), D (1.61 ppm), E (1.53 ppm), F (1.18 ppm) and G (0.86 ppm).

The olefinic proton centred on A_1 is coupled to the methyl groups D and E (long-distance coupling of type M or W) and to one of the methylene groups B.

The olefinic proton centred on A_2 is coupled to one of the methyl groups E (long-distance coupling of type M or W) and one of the methylene groups B.

The third olefinic proton A_3 is coupled to one of the methyl groups E (long-distance coupling of type M or W) and one of the methylene groups C.

There is also a coupling between the G signal at δ= 0.86 ppm and the important singlet at 1.18 ppm.

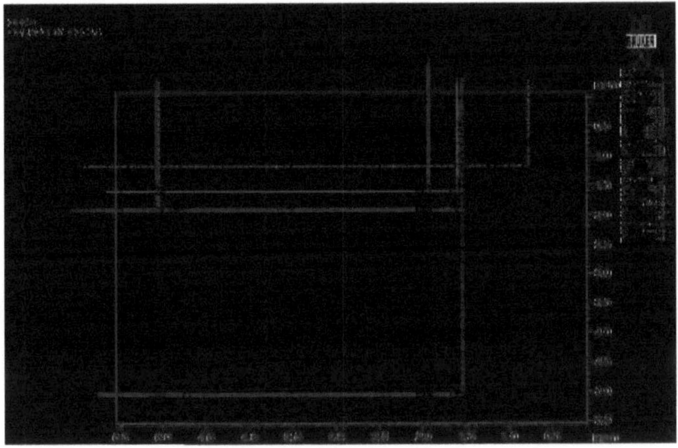

Figure 3.44. : COSY spectrum of DOR 2

Table 3.24. COSY Correlations of DOR 2

Woodpecker	δ_H (ppm)	$_H\delta$Correlations (ppm)
A1	5.03	1.61 (-CH3): A1D
		1.53 (-CH3): A1E
		1.99-2.09 (-CH2-): A1B
A2	5.05	1.53: A2E
		1.88-1.98: A2C
		1.99–2.09 : A2B
A3	5.07	1.53: A3E
		1.88-1.98: A3C
E	1.53	5.03-5.07: A1E; A2E; A3E
		1.88 - 1.98 (-CH2-): E, C
		1.99 - 2.09 (-CH2-): E, B
D	1.61	5.03 (=CH): D, A1
		1.88 - 1.98 (-CH2-): D, C
		1.99 - 2.09 (-CH2-):D, B
C	1.88 –	5.07 (=CH): C, A3
	1.98	1.53 (-CH3): C, E
B	1.99 –	5.05 (-C=CH): A2B
	2.09	1.53 (CH3): E, B
		1.61 (CH3): D, B
F	1. 18	0.86: F, G
G	0. 86	1. 18: G, F

2) HMQC Correlation

The NMR HMQC spectra (Figures 3.45 and 3.46) indicate a correlation between chemical shifts of protons and chemical shifts of the carbons to which these protons are directly attached.

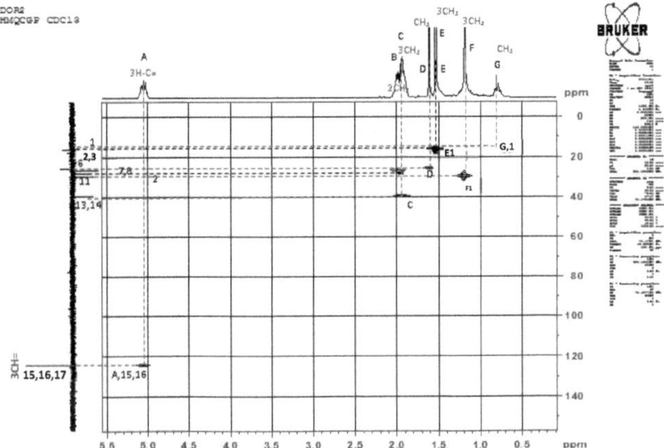

Figure 3.45. : HMQC spectrum of DOR 2

Thus, the three olefinic protons absorbing atδ: 5.03 - 5.07 ppm (A, $_{A2}$ andA3) are attached to the olefinic carbons located atδ: 124.32, 124.34 and 124.45 ppm on the spectrum.

The protons whose peak is centered atδ: 1.99 ppm (B) are coupled to the carbons ($_{CH2}$) appearing atδ: 26.70 and 26.81 ppm. The protons of the peak centered at δ: 1.96 ppm (C2 : - $_{CH2}$ -) are attached to the carbon located at δ: 28.30 ppm, and those of the peak centered at δ= 1.94 ppm ($_{C1}$: - $_{CH2}$ -) are coupled to the carbons at δ= 39.76 ppm and δ= 39.78 ppm.

The methyl group atδ: 1.61 ppm (D) is coupled to the carbon atδ: 25.71 ppm. The signal corresponding to the three methyl groups absorbing at δ: 1.53 ppm (E1, E2 and E3) correlates with carbons located at δ: 16.02, 16.07 and 13.69 ppm on the spectrum.

Finally, the protons of the peak at δ: 1.19 ppm (F) are directly attached to the carbons at δ= 29.73 ppm. ($_{CH2})_{n}$-] and the protons of the signal centered at δ= 0.86 ppm (G) are coupled to the methyl group at δ= 14.1 ppm. Table 3.25 reports these correlations.

Table 3.25. HMQC Correlations

Picks	δ_H (ppm)		Groups	Correlations (δ^{13} c)
A	5.03 – 5.07		3 = CH	124.32, 124.34, 124.45
B	1.99		2 -CH_2-	26.70, 26.81
C1	1.96		-CH_2-	28.31
C2	1.94		2 -CH_2	39.76, 39.78
D	1.61		- CH3	25.70
E	1.53		- CH3	16.02, 16.07, 13.69
F	1.18		-CH_2-	29.72
G	0.86		- CH3	14.1

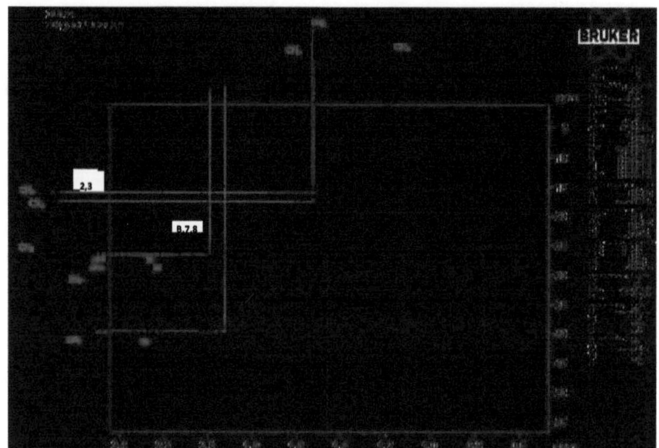

Figure 3.46. : Expansion of the HMQC spectrum of DOR 2

3) HMBC Correlations

HMBC spectra data provide information on the couplings that occur between H and C atoms separated by 2 or 3 bonds. Couplings beyond three bonds (long-distance couplings M or W) can also be observed.

Table 3.26 and Figures 3.47 and 3.48 report these correlations.

Table 3.26. : HMBC Correlations of DOR 2

Picks	δ_H (ppm)	δ^{13C} (ppm)	Groups	Correlations (δ^{13C} ppm)
A1	5.03	124.45	=C-H	
A2	5.05	124.32	=C-H	
A3	5.07	124.34	=C-H	
B	1.98 – 2.05	26.70, 26.81	- CH2 -	39.76, 39.78 (C1: ^{2J}) 124.34, 124.45, 34.92
C1	1.88 – 1.93	39.76, 39.78	- CH2 -	16.02, 16.07 (E2, E3: ^{3J}) 26.70, 26.81, (B: ^{2J}) 124.32, 124.34, 135.13
C2	1.94 – 1.98	28.31	- CH2 -	28.31, 28.31 (C2, E3: ^{2J}) 124.34 ; 135.13
D	1.61	25.70	CH3 -	13.69 (E1: ^{3J}), 25.70 (D: ^{1J}) 124.45, 131.25
E1	1.53	13.69	CH3 -	25.70 (D: ^{3J}) 124.45; 131.25
E2	1.53	16.02	CH3 -	39.76, 39.78 (C1: ^{3J}) 124.32, 134.92
E3	1.53	16.07	CH3-	39.76 (C1: ^{3J}) 124.34, 135.13
F	1.19	29.5	- (CH2)n -	29.72
G	0.86	14.5	CH3 -	31.95

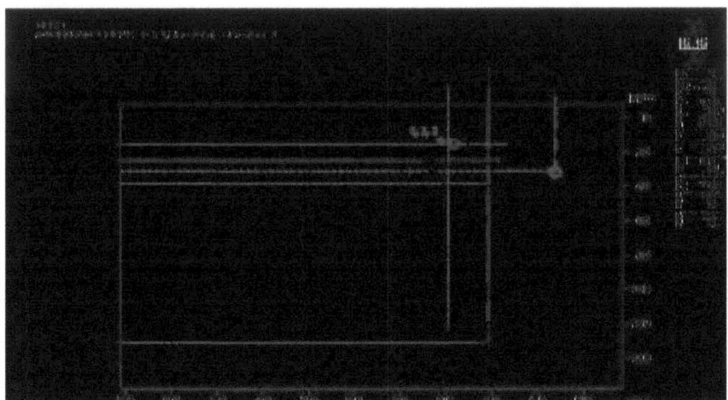

Figure 3.47. : HMBC spectrum of DOR 2

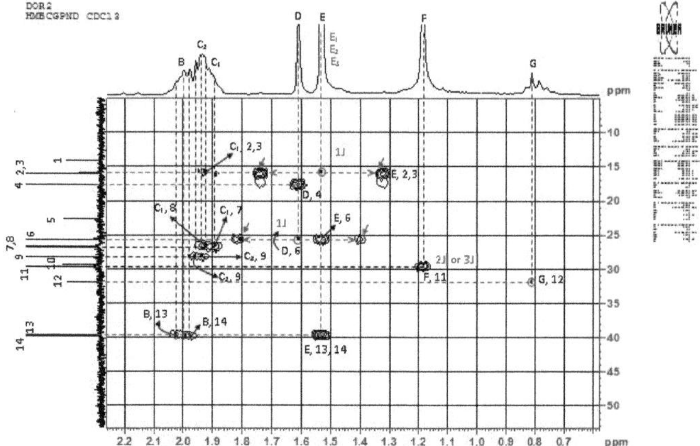

Figure 3.48. : Expansion 1 of the HMBC Spectrum of SOR 2

Thus, in HBMC-NMR (Figures 3.47 and 3.48 and Table 3.26.), the two allylic methyl groups at δ1.61 ppm (D) and 1.53 ppm (E1) are coupled with each other and with the olefinic carbon at 124.45 ppm. These methyls also present a coupling (^{2J}) with the quaternary olefinic carbon at δ131.25 ppm. This indicates that these two methyls are attached to the quaternary carbon (δ : 131.25 ppm). The latter is attached to the olefinic methine at δ: 124.45 ppm, which in turn is coupled to the methylene group at δ= 26.78 ppm (B2). In addition, the protons of these methyls present a ^{1J} coupling with the carbons to which they are respectively attached, thus confirming the HMQC data.

These data from ^{1}H, ^{13}C, DEPT-NMR and 2D-NMR correlations (COSY, HMQC and HMBC) clearly indicate that DOR 2 consists of two different compounds. The HMBC spectra show no connection between these two compounds.

1) Minority compound

With respect to the minority compound, there are characteristic data for an aliphatic hydrocarbon at δ= 0.86 ppm (m: terminal CH3) and a prominent singlet at δ= 1.18 ppm attributed to methylene groups in a [1H-NMR] aliphatic hydrocarbon [-(CH2)n-] chain (Charles *et al.*, 1993).

The [13C-NMR] data confirm these observations with the following peaks:

δ_C (ppm) :

- 14.1 (terminal CH3 of an aliphatic hydrocarbon chain)
- 22.7 (-CH2-)
- 29.4 (-CH2-)
- 29.7 (-CH2-)$_n$
- 31.9 -(CH2)

COSY data (Table 3.24) reveal that protons at δ= 0.86 ppm (signal G: CH3 terminal) are coupled to protons at = δ1.18 ppm [signal F: -(CH2)n-]. The HMQC spectrum (Table 3.25 and Figure 3.45) shows correlations between the [1H] G peak (0.86 ppm) and the signal at δ= 14.1 ppm in [13C-NMR]. There is also a coupling between the protons of the peak at δ= 1.18 ppm and the carbons at δ= 29.7 ppm in [13C-NMR]. Finally, the HMBC spectrum (Table 3.26 and Figure 3.47) shows couplings between the protons of signal G (0.86 ppm, CH3 terminal) and the methylene group carbon at 31.9 ppm. There is also a correlation between the F [-(CH2)n-] peak and the signal at δ= 29.7 ppm, attributed to the methylene groups -(CH2)n-.. These data suggest the structure below for this minority compound.

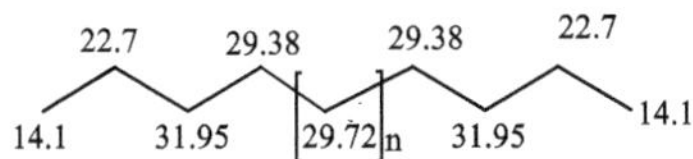

Figure 3.49. : Structure of the Minority Compound of ODR 2

2) Majority Compound

The other peaks observed in [1H-NMR], [13C-NMR], 2D-NMR belong to the second compound of DOR 2. DEPT-NMR (Figure 3.43) indicates the presence of fifteen carbon atoms including :

- Four methyl groups at δ: 16.02, 16.03. 13.69 and 25.70 ppm.
- Five methylene groups at δ: 26.70, 26.81, 28.31, 39.76 and 39.78 ppm.
- Three olefinic methines at δ= 124.32, 124.34 and 124.45 ppm
- Three olefinic quaternary carbons at δ= 131.25, 134.92 and 135.13 ppm.

This reveals the presence of three trisubstituted double bonds in the structure of this compound. The correlations between these different groups have been explained above (Tables 3.20-3.25 and Figures 3.40-3.49).

Thus, in HBMC-NMR, the two allylic methyl B groups at δ1.61 ppm (D) and 1.53 ppm (E1) are coupled with each other and with the olefinic carbon at 124.45 ppm. These methyls also present a coupling $(^{2J})$ with the quaternary olefinic carbon at δ131.25 ppm. This indicates that these two methyls are attached to the quaternary carbon (δ : 131.25 ppm). The latter is attached to the olefinic methine at δ: 124.45 ppm, which in turn is coupled to the methylene group at δ= 26.81 ppm (B_2). This methylene B_2 (δ = 26.81 ppm) is coupled to methylene C_1 (δ = 39.76 ppm).

In addition, the protons of these methyls at δ: 1.61 ppm (D) and 1.53 ppm (E1) show ^{1J} coupling with the carbons to which they are respectively attached, thus confirming the HMQC data.

This suggests the partial structure below:

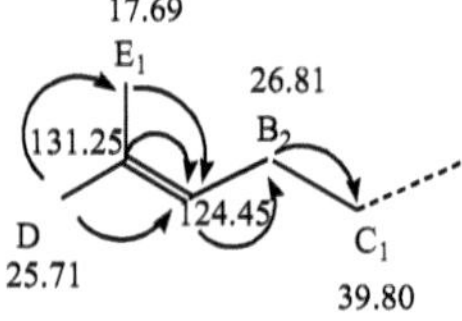

Figure 3.50. Partial Structure I of the Majority DOR 2 Compound

One of the absorbent methyls at δ= 1.53 ppm (E2) is coupled to the olefinic carbons at δ: 124.32 ppm and δ: 135.13 ppm and to the aliphatic carbon at δ: 39.8 ppm (c_1). In addition,

116

the two -CH2- groups appearing at δ= 1.98 - 2.05 ppm are coupled with olefinic methines at : 124.832 ppm (B1, 15) and δ= 124.45 ppm (B2, 17). One of these methylenes (B1) correlates with the quaternary carbon at δ= 135.13 ppm (B1, 19), suggesting the partial structure :

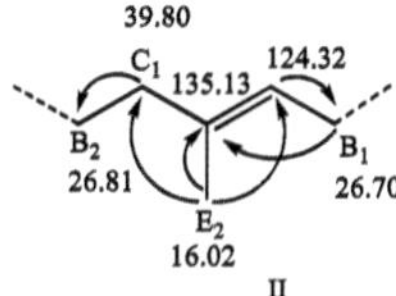

Figure 3.51.: Partial structure II of the majority SOR 2 compound

Finally, the peak of the two -CH2 - centred atδ: 1.88 -1.95 ppm (c1) shows couplings with the methyl groups absorbing at δ= 16.02 ppm (C1,E2) and 16.07 (c1,E3). Each of these methylene groups is also coupled to methylenes absorbing at δ= 26.70 ppm (B1,C1) and 26.81 ppm (B2,C1). They are each coupled to the olefinic methine groups absorbing atδ: 124.32 ppm (c1,15), and 124.34 ppm (c1,16). In addition, one is coupled to the olefinic quaternary carbon atδ: 134.92 ppm (c1,19) and the other is coupled to the olefinic quaternary carbon at δ= 135.12 ppm (c1,20).

This suggests the following partial structures:

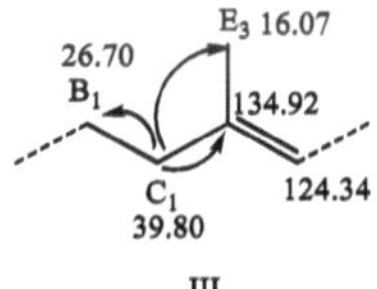

Figure 3.52. Partial Structure III of the Majority DOR 2 Compound

The two methylenes absorbing atδ: 1.94 - 1.98 ppm (c2) correlate with olefinic methine at δ= 124.34 ppm and with olefinic quaternary carbon at δ= 134.92 ppm. In addition, they are coupled to each other (c2, 9 and c2, 9).

This suggests the partial structure below:

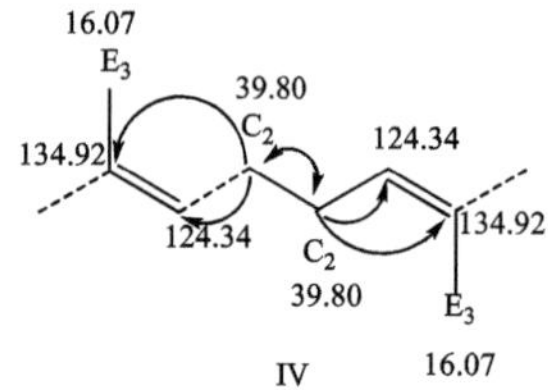

Figure 3.53. : Partial structure IV of the majority compound of DOR 2

By connecting these partial structures, we obtain this intermediate structure:

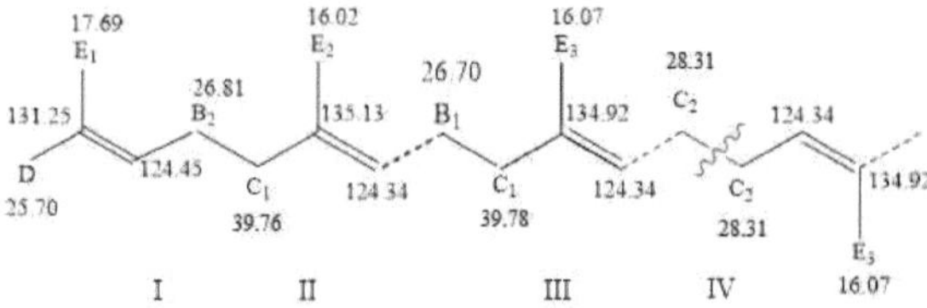

Figure 3.54. : Connection of 4 partial structures

The HMBC-NMR data therefore suggest the structure of a symmetrical unsaturated hydrocarbon having eight allylic methyl groups and six double bonds. The plane of symmetry passes through the c_2-c_2 bond of the interim structure. Thus, given the symmetrical nature of the molecule, the integration at [1H-NMR] and the number of carbon atoms at [13C-NMR] indicates only half of the molecule. Therefore, the molecular formula deduced from the NMR-DEPT data is C30H50. All [1H], [13C-NMR] data and all 2D-NMR correlations identify this compound with a known triterpene: squalene. They are comparable to those in the literature (Mudiyanselage *et al.*, 2003; Borchman *et al.*, 2013).

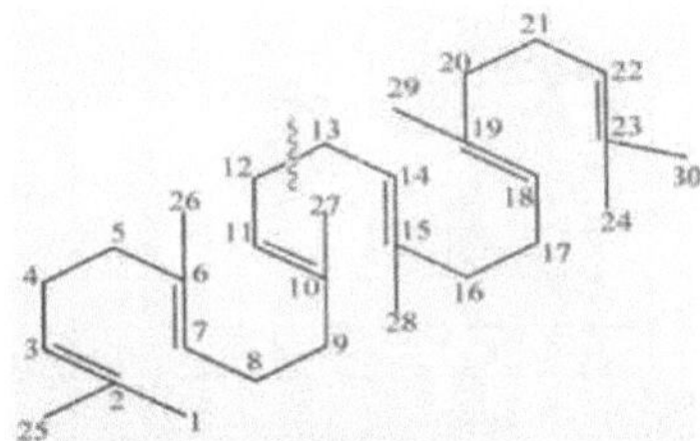

Figure 3.55.: Squalene

These data also identify the minority compound of DOR 2 to the aliphatic hydrocarbon n-undecane (Charles *et al.*, 1993).

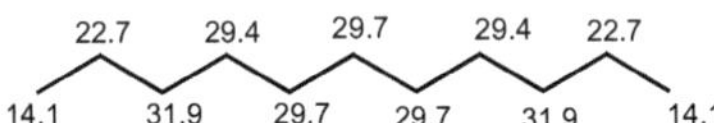

Figure 3.56. *n-Undecane*

3.1.5.6. Analysis of the Tsh-Do-4-3-II   sample from the ethyl acetate extract

The ¹H-NMR spectrum (Figure 4.20 in the Appendix and Table 3.27.) indicates that **Tsh-Do 4-3-II is** also a product mixture in which the majority product is a long-chain saturated hydrocarbon. It is contaminated with an aromatic compound and glycerides.

The ¹³C-NMR spectrum (Figure 4.21 in Appendix and Table 3.28.) confirms the presence of a majority hydrocarbon.

Table 3.27. [1H-NMR] spectrum of M-TSD-DO-4-3-II

Woodpecker	δ_H	Multiplicity	Group		δ_H	Multiplicity	Group
1	*0.88*	t	***Terminal***	12	4.31	m	
2	1.10	s		13	4.96	q	-CH2 -
3	*1.26*	bs	*-(CH2)n-*	14	5.82	m	
4	1.61			15	6.51		
5	1.70			16	6.99	s	
6	2.04	q		17	3.36		
7	2.17	s		18	3.46	s	
8	2.59	m		19	3.49	s	
9	2.87	t		20	3.53	m	
10	3.63	m		21	3.71	m	
11	4.07	m		22	3.80	s	
				23	3.93	s	

The peaks in bold italics correspond to peaks in the aliphatic chain. The other peaks belong to the contaminants.

Table 3.28. [13C-NMR] spectrum of TSh-DO-4-3-II

Woodpecke	δ_C (ppm)	Group
1	14.1	***-CH3***
2	22.6	-CH2 -CH3
3	29.3	-CH2 -
4	29.6	*-(CH2)n-*
5	31.9	-CH2-CH2 -CH3

Although the [1H-NMR] and [13C-NMR] data alone are not sufficient to draw an irrefutable conclusion, one can, based on [1H-NMR] integration, suggest that the hydrocarbon could be identified as ***n-eicosane***: CH3-CH2)18-CH3 (Charles *et al.*, 1993).

From the above it is noted that the methanolic extract of these three plants has almost the same chemical profile. Indeed, one notes the presence of phenolic compounds such as **flavonoids of the kæmpferol type,** phenolic acids in this case **rosmarinic acid**. The latter is largely in *Ocimum basilicum* followed by *O. gratissimum and* finally *O. canum* which, in turn, has another compound with a retention time equal to 58,719 min on the HPLC chromatogram (Figure 3.2). This compound is in trace amounts in the other two species (Tshilanda *et al.,* 2016b).

Chemical screening on the three plants confirms this similarity but the sensitivity to certain reagents is greater for *Ocimum gratissimum* than for the other two species. In addition Ocimum *canum* did not react to the moss test which indicates the presence of saponins in a plant.

Chromatographic and spectroscopic analyses carried out on three of four extracts from the solvents with increasing polarity of *Ocimum gratissimum*, made it possible to isolate and identify at least **35** compounds whose names and extract of origin are listed below.

1) **Extract ED:**

Squalene, n-tridecane, oleanolic acid, trans-caryophyllene, β-bisabolene, epi-α-selinene, *n-tetradecane,* cyclohexatetracarontane, heneicosane, nonadecane, nonacosane, (Z)-5-undecen-2-ol, (*E*)-5-undecen-2-ol, heptadecane, 2-phenyl-2-tipyl acenapthenone, 2(*E*)-*nonadecene,* 3(*E*)-eicosene, dioctadecyl phosphonate.

2) **AcoEt extract:** *n-eicosane*, ursolic acid, oleanolic acid and palmitate from a triterpenoid.

3) **MeOH extract:** 1,8-dimethoxy-3-methyl-3-anthraquinone, triacontane, hexatriacontane,

n-Eicosane, 2,6-di-tert-butylphenol, n-heptadecane, α-octadecene, 1-heptadecene, 2-phenyl-2-tipyl-acenapthenone, tritetracontane, tetradecane, benzene dibutyl-1,2-dicarboxylate or butylphthalate, docosene-1, methyl hexadecanoate, eicosanol-1, bis (2-ethylhexyl) or bis (2-ethylhexyl) phthalate benzene-1,2-dicarboxylate, diterbutylphenol and cyclohexatetracarontane.

Butyl stearate has been isolated from *Ocimum basilicum* (Tshilanda et *al.*, 2014) and currently the compound named **"Do8"**.

NMR data for **Tsh-Do-1-2-II, Tsh-DO-2-2-II** from the DCM extract of *Ocimum gratissimum* and near-pure **Do8 from the** acidified methanol extract of *Ocimum basilicum were not* sufficient to elucidate the chemical structures of their constituents. Their analysis will be continued at a later stage.

Since the genus *Ocimum* includes aromatic plants, a study on the extraction of the essential oils of our three species, the analysis by GC and GC/MS of these essential oils to determine their chemical composition was carried out.

3.1.6 Extraction and analysis of essential oils

3.1.6.1. Extraction of essential oils

One hundred grams of leaves of *Ocimum basilicum* gave after hydrodistillation 0.648 g of essential oil, i.e. a yield of 0.65%, while 360 g of leaves and inflorescences of *Ocimum canum* Sims gave after hydrodistillation 2.188 g of essential oil, i.e. a yield of 0.61%. As for *Ocimum gratissimum,* 150 g of leaves and inflorescences gave, after hydrodistillation, 0.4894 g of essential oil, i.e. a yield of 0.326%.

One millilitre of each oil was weighed to determine its density (specific gravity).

Table 3.29 below shows the yield in mass percentage of essential oil for each species studied and its density.

Table 3.29 : Percentage return of three *Ocimum*

Plant species	Yield in % of essential oil	Density
Ocimum basilicum	0,65	0,918
Ocimum canum	0,61	0,932
Ocimum gratissimum	0,33	0,804

Ocimum basilicum and *Ocimum canum* have almost the same content of volatile constituents (0.65% and 0.61% respectively) while the yield of essential oil from *Ocimum gratissimum* is almost half of two others.

It is known that the essential oil content depends on the season. This is the case of the essential oil of *Ocimum basilicum* from Pakistan whose content varies from 0.5% to 0.8% with a maximum   content for the winter harvest while a minimum content has been observed in summer (Hussain et *al.*, 2008).

Studies on *Ocimum basilicum from* western Lafayette and Australia have shown higher essential oil yields than ours (case of *O. basilicum*) (Denys & James, 1990; Katarzyna et *al.*, 1996), others have shown lower yields (0.07%) for some varieties of Ocimum *basilicum* harvested in Mississippi (Valtcho et *al.*, 2008).

It is also noted that each oil has a different density from the others. The essential oil of *O. canum* is the densest of the three while that of *O. gratissimum is the* least dense.

3.1.6.2. Analysis of essential oils of three species of Ocimum

a) Ocimum canum

The GC/MS chromatogram of *Ocimum canum* Sims leaf oil (Figure 3.57) shows peaks showing the presence of nine components, one of which is the majority component (given the intensity of its peak) and which was identified at 1.8 kineol (TR=7,653min) by comparison of its mass spectrum with its counterpart in the Wiley 275 database (see Fig. 4.22 in the Appendix).

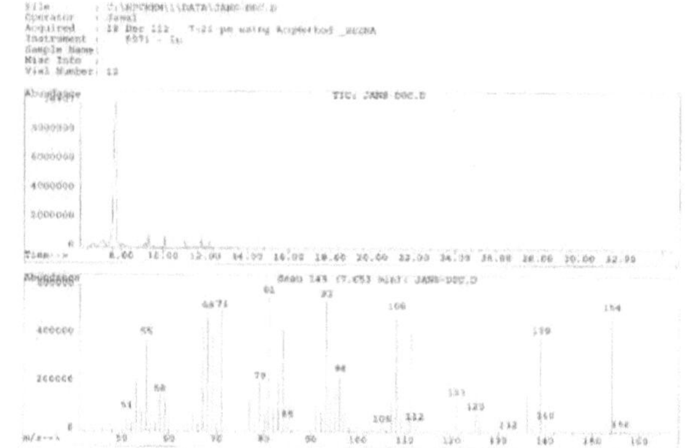

Figure 3.57: Chromatogram of *Ocimum canum* Sims (DoC) leaf essential oil and the mass spectrum of the peak-intense compound

This result confirms the 1,8-cineole chemotype for *Ocimum canum from* DRC which is the same as the one found for the essential oil of *Ocimum canum leaves from* Burkina-Faso where 1,8-cineole is 61.2% and approaches that of *O leaves.canum from* one of two sources on the Island of Grande Comore which is 48% 1,8-cineole followed by camphor (14.99%), α -pinene (5.71%) β- pinene (4.66%) (Bassole *et al.* , 2001; Hassan et *al.* , 2011).

On the other hand, the essential oil of *Ocimum canum from* Rwanda and Cameroon is characterized by linalool at 56.3% and 44.9% respectively (Weaver et *al.,* 1991; Tchoumbougnang *et al.,* 2008), testifying to the linalool chemotype.

(b) Ocimum basilicum

The GC/MS chromatogram of the essential oil of *Ocimum basilicum* (Figure 3.58) shows peaks corresponding to 40 compounds of which 35 have been identified and represent 98.1% of the total essential oil. Figures 3.59, 3.60 and 3.61 represent the different parts of

Figure 3.58. Figure 3.59 has been amplified to better illustrate the minority constituents of the first part of the chromatogram (Figure 3.58).

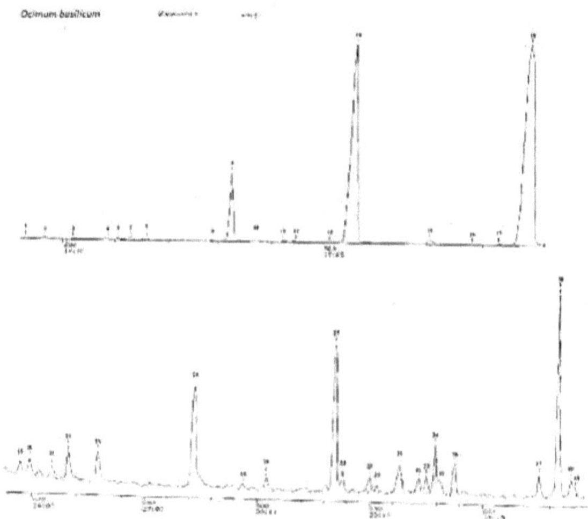

Figure 3.58. Chromatogram of the essential oil of *Ocimum basilicum:* all constituents

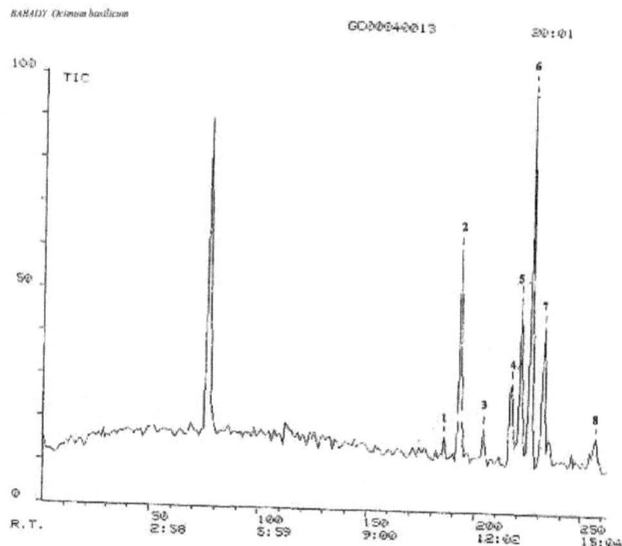

Figure 3.59. : Chromatogram of the essential oil of *Ocimum basilicum*: minority constituents (1-8)

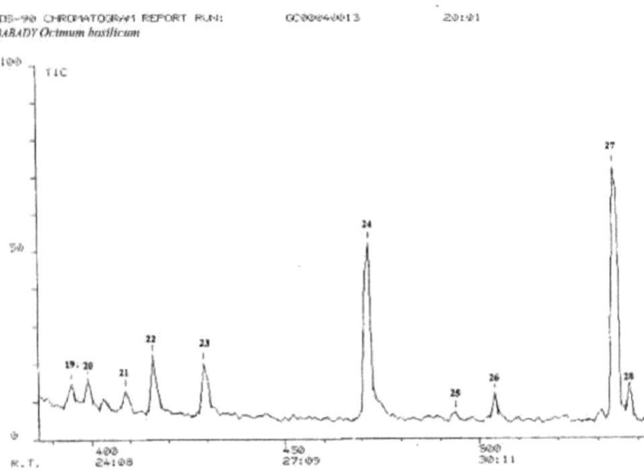

Figure 3.60.: Chromatogram of the essential oil of *Ocimum basilicum*: minority constituents (19-28).

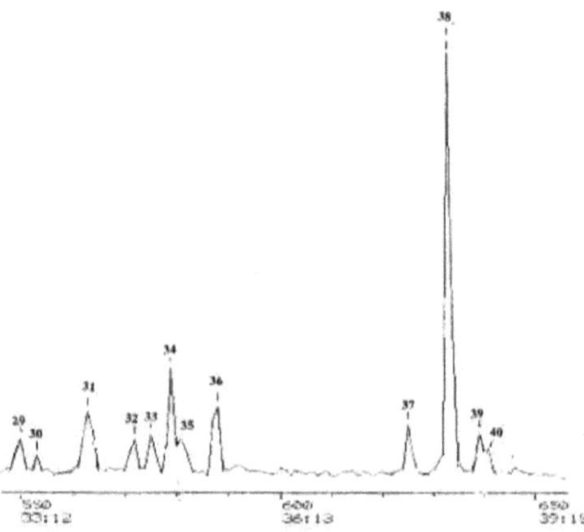

Figure 3.61. Chromatogram of the essential oil of *Ocimum basilicum:* minority constituents (29-40)

The volatile constituents of this essential oil (Table 3.30) can be classified into three categories:

- the majority components: the concentration is higher than 1.5%.
- Minority components: the concentration is between 0.05- 1.5%.
- Trace components: The concentration is less than 0.01%.

The identification of compounds was made possible by comparing their mass spectra (MS) and retention indices (RI) with those in the database; the retention index is used to convert retention times into system-independent constants.

Table 3.30. Chemical composition of the essential oil extracted from the leaves of *Ocimum basilicum*

Peak number	Compound	IRa	IRb	%	Methods of identification
1	α-Thujène	921	928	0,02	SM, IR
2	α-Pinene	931	932	0,13	SM, IR
3	Camphene	942	946	0,04	SM, IR
4	Oct-1-in-3-ol	965	964	0,04	SM, IR
5	Sabinène	966	968	0,03	SM, IR
6	β-Pinene	970	972	0,20	SM, IR
7	Myrcene	982	984	0,03	SM, IR
8	p-Cym	1013	1015	0,02	SM, IR
9	**1.8-Cineole**	**1020**	**1023**	**4,04**	SM, IR
10	trans-Ocimene	1035	1038	0,16	SM, IR
11	γ-Terpinene	1048	1050	0,02	SM, IR
12	Cis-sabinene hydrate	1083	1054	0,02	SM, IR
13	Not identified (C10H12O)	-	-	0,01	SM: m/z: 44, 59, 65, 91(100%), 118, 120, 131
14	**Linalol**	**1084**	**1085**	**21,25**	SM, IR
15	Camphor	1126	1125	0,23	SM, IR
16	δ-Terpineol	1143	1148	0,02	SM, IR
17	Terpineol-4	1161	1164	0,02	SM, IR
18	**Estragole**	**1173**	**1178**	**35,72**	SM, IR
19	Nérol	1213	1216	1,25	SM, IR
20	(Z)-Citral	1217	1220	1,40	SM, IR
21	Geraniol	1235	1237	1,01	SM, IR
22	**(E)-Citral**	**1242**	**1247**	**1,55**	SM, IR
23	**Thymol**	**1275**	**1272**	**1,64**	SM, IR
24	**Eugenol**	**1335**	**1340**	**4,60**	SM, IR
25	Not identified: mixture of two compounds (C12H18O + ?)	-	-	0,23	SM: m/z: 41, 48, 56, 66, 79, 121, 137, 146, 164, 178(100%), 187, 195
26	α-Copaene	1372	1374	0,53	SM, IR
27	*cis-α-Bergamotene*	**1408**	**1410**	**6,56**	SM, IR
28	*cis-Caryophyllene*	1415	1419	0,61	SM, IR
29	α-Humulene	1451	1450	0,85	SM, IR
30	Unidentified: (C15H24O)	-	-	0,15	SM: m/z: 53, 80, 81, 91, 105, 117, 119, 131, 137, 147, 161 (100%)
31	**Alloaromadendrene**	1452	1459	1,30	SM, IR
32	Not identified (C15H24O)	-	-	0,65	SM: m/z: 41, 53, 66, 69, 79, 81, 91(100%), 93, 107, 119, 121, 133, 147, 161, 175, 204
33	Not identified (C15H24O)	-	-	0,84	SM: m/z: 41, 55, 67, 79,81, 91(100%), 93, 105, 107, 108, 119, 121, 135, 139, 147, 148, 161, 167, 189, 204
34	**Germacrene D**	**1471**	**1476**	**2,06**	SM, IR
35	δ-Cadinene	1515	1514	1,23	SM, IR
36	(E)-α-Bisabolene	1532	1534	1,27	SM, IR
37	1,10-di-epi-Cubenol	1612	1606	0,95	SM, IR

38	**epi-α-Cadinol**	**1622**	**1626**	**8,02**	SM, IR
39	β- Eudesmol	1630	1631	0,72	SM, IR
40	α-Eudesmol	1637	1641	0,52	SM, IR

Constituents are presented in column elution order, IRa: retention indices measured on apolar column (OV-1), IRb: literature retention indices (Babushok et *al.,* 2011; Bendiabdellah *et al.,* 2012).

The total of constituents is 99.98% distributed as follows:

- Monoterpenes: 0.67%.

- Oxygenated monoterpenes: 26.75%.

- Non-oxygenated sesquiterpenes: 14,41 %.

- Oxygenated sesquiterpenes: 10.21%.

- Oxygenated hydrocarbons: 0.04%.

- Aromatic hydrocarbons: 0,02 %.

- Oxygenated aromatic hydrocarbons: 46,00 %.

- Unidentified compounds: 1,88 %.

The major compounds are methylchavicol or estragole (35.72%), linalool (21.25%), epi-α-cadinol (8.02%), α-bergamotene (6.56%), eugenol (4.60%), 1,8-cineole (4.04%), germacrene D (2.06%), thymol (1.64%), (*E*)-citral (1.55%). They represent 84.44%) of the total concentration of the essential oil.

The minority compounds are : (*Z*)-citral (1.40%), alloaromadendrene (1.30%), *(E)*-α-bisabolene (1.27%), δ-cadinene (1.23%), nerol (1.25%), geraniol (1.01%), 1,10-di-epi-cubenol (0, 5%), α-humulene (0.85%), β-eudesmol (0.72%), *cis-caryophyllene (0.*61%), α-copaene (0.53%), α-eudesmol (0.52%), camphor (0.23%), β-pinene (0,20%), trans-ocimene (0.16%), α-pinene (0.13%), camphene (0.04%), oct-1-en-3-ol (0.04%), sabinene (0.03%), myrcene (0.03%), α-thujene (0.02%),   p-cymene (0.02%), γ-terpinene (0.02%), *cis-sabinene* hydrate (0.02%), δ-terpineol (0.02%), terpineol-4 (0.02%) and five unidentified constituents.

These results indicate that the essential oil of *Ocimum basilicum* is rich in oxygenated aromatic hydrocarbons (46.00%) and oxygenated monoterpenes (26.75%). The most abundant constituent of this essential oil is methylchavicol (35.72%), followed by linalool (21.25%).

Oxygenated sesquiterpenes and non-oxygenated sesquiterpenes represent respectively 14.41% and 10.21% of the total essential oil. Non-oxygenated monoterpenes account for only 0.67 %, unidentified compounds represent 1.88 % and non-oxygenated aromatic hydrocarbons represent 0.02 % of the total content of this essential oil.

This analysis highlights the fact that the essential oil of *Ocimum basilicum* L. from the DRC has a composition richer in oxygenated compounds than in terpene hydrocarbon compounds. Similar studies on the assessment of the chemical composition of the essential oil of *O. basilicum L from* other countries have reported the presence of a few compounds found in the essential oil of *O. basilicum L from* DRC and for the most part linalool is the major component (James *et.* al., 1999; Labra *et al.,* 2004; Lee et al., 2005; Chalchat & Ozcan, 2008; Klimankova et al., 2008; Dabolena et al., 2010).

Although the composition of the essential oil changes with the origin of the plant and evolves according to the production conditions, the   methyl-chavicol-rich composition of the essential oil of the DRC species is typical of the essential oil of *O. basilicum* tropical, and is identical to that of the Indian species, whereas that of European basil is of the chemotype linalol or methyl eugenol (Ozcan & Chalchat, 2002; Zhang et *al.* , 2009; Ismail et *al.* , 2010; Dambolena et al.., 2010).

Of the seven chemotypes listed for *Ocimum basilicum* L based on the chemical composition of essential oils (Valtcho et *al.,* 2008), it is noted that our species displays the methyl chavicol-linalol chemotype.

C) Ocimum gratissimum

The GC/MS chromatograms of the essential oil of leaves of *Ocimum gratissimum* (Figures 3.62-3.64) show peaks corresponding to the majority compounds: 45 compounds, 38 of which represent 99.58% of the total essential oil, have been identified.

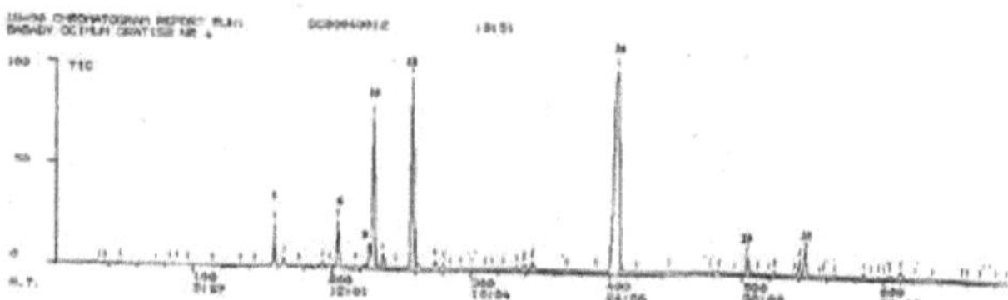

Figure 3.62. : Chromatogram of the essential oil of *Ocimum gratissimum:* majority constituents

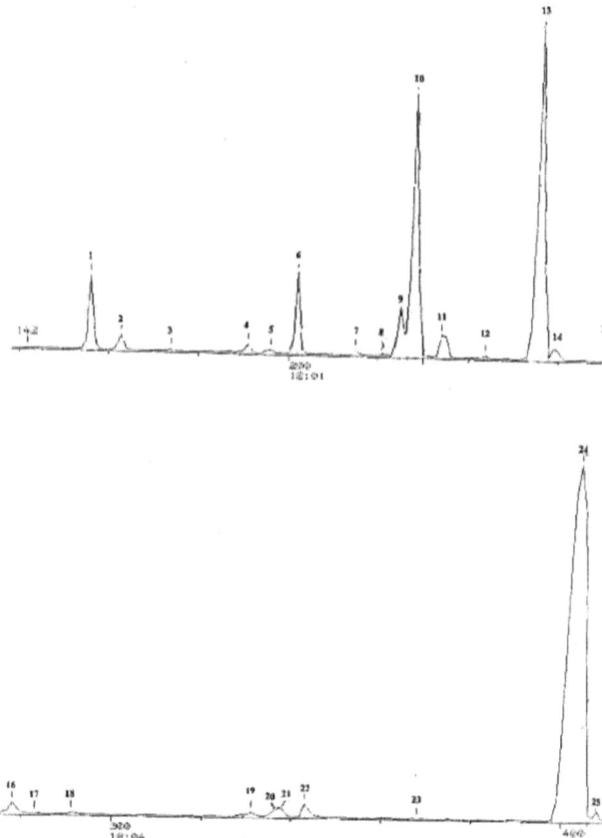

Figure 3.63. Chromatogram of the essential oil of *Ocimum gratissimum*: compounds (1-25).

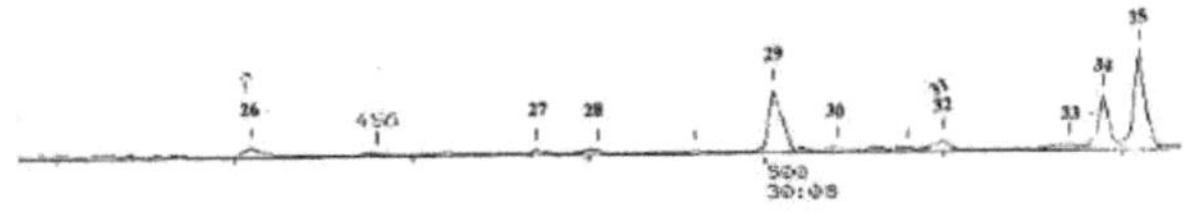

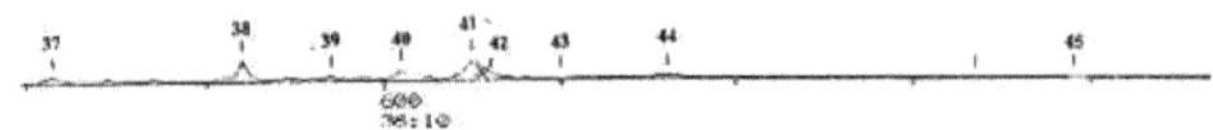

Figure 3.64. Chromatogram of the essential oil of *Ocimum gratissimum*: compounds (26-45).

The volatile constituents of this essential oil identified by comparison of their mass spectra (MS) and retention indices (RI) with those of their counterparts in the database are shown in Table 3.31.

Table 3.31. Chemical composition of the essential oil extracted from the leaves of *Ocimum gratissimum*

Peak No.	Compound	IRa	IRb	%	Methods of identification
1	**α-Thujène**	**923**	**926**	**2,79**	SM, IR
2	α-Pinene	931	934	0,59	SM, IR
3	Camphene	944	947	0,05	SM, IR
4	Sabinène	964	968	0,41	SM, IR
5	β-Pinene	970	973	0,20	SM, IR
6	**Myrcene**	**981**	**983**	**2,79**	SM, IR
7	α-Phellandrene	998	1000	0,10	SM, IR
8	Δ3-Hull	1005	1007	0,05	SM, IR
9	α-Terpinene	**1009**	**1011**	**2,23**	SM, IR
10	**p-Cym**	**1012**	**1015**	**17,60**	SM, IR
11	Sylvestrène	1022	1024	0,91	SM, IR
12	*(E)*-β-Ocimene	1036	1038	0,07	SM, IR
13	**γ-Terpinene**	**1048**	**1050**	**22,39**	SM, IR
14	*cis-Sabinene* hydrate	1053	1056	0,47	SM, IR
15	p-Cymenen	1072	1074	0,53	SM, IR
16	Linalol	1083	1086	0,47	SM, IR
17	trans-Sabinene hydrate	1085	1087	0,09	SM, IR
18	Unidentified	-		0,08	SM: m/z: 41(100%), 67, 81,109,
19	Unidentified	-	-	0,24	SM: 41.79, 91, 107(100%), 121,
20	p-Cymen-8-ol	1160	1162	0,27	SM, IR
21	Terpinen-4-ol	1161	1164	0,68	SM, IR
22	Estragole	1179	1177	0,51	SM, IR
23	Thymyl methyl ether	1212	1214	0,03	SM, IR
24	**Thymol**	**1275**	**1272**	**39,86**	SM, IR
25	Carvacrol	1279	1283	0,20	SM, IR
26	Eugenol	1336	1340	0,24	SM, IR
27	α-Copaene	1377	1376	0,06	SM, IR
28	2,3-Dimethoxy-p-	1409	1407	0,09	SM, IR
29	**(E)-β-Caryophyllene**	**1417**	**1419**	**1,54**	SM, IR
30	*(E)*-β-Farnesene	1445	1444	0,05	SM, IR
31	*(Z)*-tert-butyl-hydroxyanisole?	1448	?	tr	SM, IR
32	α-Humulene	1451	1449	0,20	SM, IR
33	Alloaromadendrene	1456	1459	0,13	SM, IR
34	β-Sélinène	1482	1480	0,81	SM, IR
35	**β -Bisabolene**	**1496**	**1500**	**1,86**	SM, IR

36	7-epi-α-Selinene	1522	1508	0,09	SM, IR
37	Elémol	1535	1536	0,13	SM, IR
38	Caryophyllene oxide	1562	1570	0,42	SM, IR
39	Humulene oxide I	1593	1591	0,07	SM, IR
40	γ-Eudesmol	1612	1617	0,24	SM, IR
41	β-Eudesmol	1629	1634	0,36	SM, IR
42	α-Eudesmol	1639	1641	tr	SM, IR
43	Unidentified	-	-	tr	SM: 79, 81, 91, 105 (100%),
44	Unidentified	-	-	0,05	SM: 39, 67, 81, 91, 105 (100%)
45	Unidentified	-	-	tr	SM: 41, 69, 81(100%), 93, 121,

Constituents are presented in column elution order, IRa: retention indices measured on apolar column (OV-1), IRb: literature retention indices (Babushok *et al.*, 2011; Bendiabdellah *et al.*, 2012).

The total content of all the components of this essential oil is 99,95 % distributed as follows :

- Non-oxygenated monoterpenes: 32.58%.
- Oxygenated monoterpenes: 1.71%.
- Non-oxygenated sesquiterpenes: 4.49%.
- Oxygenated sesquiterpenes: 1.22%.
- Aromatic hydrocarbons: 18.13%.
- Oxygenated aromatic hydrocarbons: 41.20%.
- Unidentified: 0.37

The majority compounds are thymol (39.86%), γ-terpinene (22.39%), p-cymene (17.60%), α-thujene (2.79%), myrcene (2.79%), α-terpinene (2.23%), β-bisabolene (1.86%) and (*E*)-β-caryophyllene (1.54%). They represent 91.06% of the total concentration of the essential oil.

The minority compounds are : α-pinene, camphene, sabinene, β-pinene, sylvestrene, α-phellandrene, Δ3-carene, *(E)*-β-ocimene, *cis-sabinene* hydrate, p-cymenene, linalool, trans-sabinene hydrate, p-cymen-8-ol, terpinen-4-ol, estragole, thymylmethyl ether, eugenol, α-copaene, 2,3-dimethoxy-p-cymene, *(E)*-β-farnesene, α-humulene, alloaromadendrene, β-selinene, 7-epi-α-selinene, elemol, caryophyllene oxide, humulene epoxide, γ-eudesmol, β-eudesmol, α-eudesmol, and three unidentified constituents.

The trace compounds present are: *(Z)*-tert-butyl-hydroxyanisole, α-eudesmol and two unidentified compounds.

The essential oil ofOcimum *gratissimum* is rich in oxygenated aromatic hydrocarbons (41.20%) and monoterpenes (32.58%). The most abundant compound of this essential oil is thymol (39.86%). Aromatic hydrocarbons account for 18.13% of the essential oil while oxygenated monoterpenes, sesquiterpenes and oxygenated sesquiterpenes account for only 1.71%, 4.49%, and 1.22% respectively. Unidentified compounds account for 0.37% of this gasoline.

Our sample contains a rare isomer of limonene, sylvestrene. This is the first time this constituent has been found in the essential oil of *Ocimum gratissimum*. This essential oil contains (Z)-tert-butyl-4-hydroxyanisole, a powerful antioxidant used in the food, cosmetic and pharmaceutical industries for the protection of oils and fats. It is generally obtained by organic synthesis as a mixture with its isomer, 3-tert-butyl-4-hydroxyanisole. Owokotome et *al* (2012) reported for the first time the presence of these two isomers in the essential oil of *Ocimum gratissimum* harvested in Nigeria and recognized as a natural source for food preservation.

The three major components of the essential oil of *Ocimum gratissimum from* DR Congo are comparable to those of *Ocimum gratissimum* from Benin but differ in their content and order of magnitude. For the Benin sample, the thymol content is 43.4% followed by p-cymene (12.3%) and γ-terpinene (12.1%) (Kpodekon et *al.*, 2013) while for our sample (sample T from DR Congo) the thymol content is 39.0% followed by γ-terpinene (22.39%) and p-cymene (17.60%).

The essential oils of *Ocimum gratissimum*, *O. basilicum* and *O. canum* have already been the subject of several studies in different countries, but this is the first time that a study has been done on the essential oil of *Ocimum canum* growing in Mbuji-Mayi in the Democratic Republic of Congo.

In this work, an extensive study of the composition of essential oils of two other species growing in DR Congo was undertaken: *Ocimum gratissimum* L and *Ocimum basilicum* L although their essential oils have already been analysed by Cimanga and colleagues (Cimanga *et al.*, 2002).
Comparison of their results with ours indicates some similarities but also some important differences between the two samples.

Thus, in the case of *Ocimum gratissimum*, 16 constituents were identified in the Cimanga sample (C sample) while 38 constituents were identified in the Tshilanda sample (T sample). Twelve compounds are the same in both samples but their concentrations are different as shown in Table 3.32. Four compounds identified in the C sample are absent from the T sample: limonene, 1,8-cineole, α-terpineol, and β-elemene. Twenty-eight constituents identified in our essential oil (Sample T) were not detected in Sample C. Table 3.32 below gives all this comparison.

Table 3.32. Comparison of the chemical composition of essential oils of *Ocimum gratissimum* L. (Cimanga Sample and Tshilanda Sample)

% (Cimanga Sample: C)	Constituent	% (Tshilanda Sample: T)
0,5	**α-thujene**	**2,79**
2,8	α-pinene	0,59
-	camphene	0,05
-	Sabinène	0,41
0,6	β-pinene	0,20
0,3	**Myrcene**	**2,79**
0,4	α-phellandrene	0,10
-	Δ3 hull	0,05
-	**4α-terpinene**	**2,23**
7,3	**p-cymene**	**17,60**
2,3	Limonene	-
	Sylvestrène	0.91
3,4	*(E)*-β-Ocimene	0.07
25,7	**γ-terpinene**	**22,39**
-	*cis-sabinene* hydrate	0,47
-	p-cymenene	0,53
-	Linalol	0,47
-	trans-sabinene hydrate	0,09
	Unidentified	0,8
	Unidentified	0,24
-	p-cymen-8-ol	0,27
1,5	Terpine 4-ol	0,68
-	Estragole	0,51
-	Thymyl methyl ether	0,03
53,2	**Thymol**	**39,86**
1,60	**Carvacrol**	0,20
12,7	**Eugenol**	0,24
-	α-copaene	0,06
-	2,3-dimethoxy-p-cymene	0,09
-	*(E)*-β-caryophyllene	**1,54**

	(E)-β-farnesene	0,05
-	(Z)-tert-butyl-4-hyroxyanisole	0,04
-	α-humulene	0,20
-	Alloaromadendrene	0,13
-	β-selinene	0,81
-	**β -bisabolene**	**1,86**
-	7-epi-α-selinene	0,09
-	Elémol	0,13
-	Caryophyllene oxide	0,42
-	Humulene oxide I	0,07
-	γ-eudesmol	0,24
-	β-eudesmol	0,76
-	α-eudesmol	tr
-	Unidentified	tr
-	Unidentified	0,05
-	Unidentified	tr
0,5	1.8-cineole	-
0,2	α-terpineol	-
0,1	β-élémène	-

The major constituents are thymol, γ-terpinene, eugenol and p-cymene in the C sample and thymol, γ-terpinene, and p-cymene in the T sample, but eugenol is a minority constituent in the T sample. The most abundant compound in both samples is thymol but concentrations vary by sample: 53.20% in sample C and 39.86% in sample T. The concentrations of the other majority compounds are also different. Thus, the content of γ-terpinene is 25.7% (sample C) and 22.4% (sample T); the difference in p-cymene content is very significant: 17.6% in the T sample while it is only 7.3% in the C sample.

For *Ocimum basilicum*, 13 constituents were identified in the C sample while 35 constituents were identified in the T sample. Eleven products are the same in both samples but the concentrations are different as shown in Table 3.39. Two constituents identified in the C sample are absent from the T sample: limonene and β-elemene. Twenty-four constituents identified in the T-sample were not detected in the C-sample as further reported in Table 3.33.

Table 3.33. Comparison of the chemical composition of essential oils of *Ocimum basilicum* (Cimanga Sample and Tshilanda Sample)

% (Cimanga Sample)	Compound	% (Tshilanda sample)

0,2	α-thujene	0,02
2,6	**α-pinene**	0,13
-	camphene	0,04
-	Oct-1-en-3-ol	0,04
-	Sabinène	0,03
0,5	β-pinene	0,20
0,2	Myrcene	0,03
3,2	**p-Cym**	0,02
3,2	**1.8-Cineole**	**4,04**
-	trans-Ocimene	0,16
1,2	γ-terpinene	0,02
-	Cis-sabinene hydrate	0,02
-	Unidentified	0,01
-	**Linalol**	**21,25**
-	Camphor	0,23
-	δ-terpineol	0,02
1,4	Terpineol-4	0,02
-	**Estragole**	**35,72**
-	Nérol	1,25
-	(Z)-citral	1,40
-	Geraniol	1,01
-	(E)-citral	1,55
43,5	Thymol	1,64
8,1	**Eugenol**	**4,60**
-	Unidentified	0,23
-	α-copaene	0,53
-	**α-bergamotene**	**6,56**
-	*cis-caryophyllene*	0,61
0,6	α-humulene	0,85
-	Unidentified	0,15
-	Alloaromadendrene	1,30
-	Unidentified	0,65
-	Unidentified	0,84
-	**Germacrene D**	**2,06**
-	δ-cadinene	1,23
-	(E)-α-bisabolene	1,27
-	1,10-di-epi-cubenol	0,95
-	**epi-α-cadinol**	**8,02**
-	β- eudesmol	0,72
-	α-eudesmol	0,52
0,2	β-élémène	-

2,4	limonene	-

The major constituents are thymol and eugenol in sample C while in sample T estragole (methylchavicol), linalool, epi-α-cadinol, α-bergamotene, eugenol, and 1,8-cineole are major constituents. Thymol is also a major constituent in the T-sample, but its concentration is only 1.64%. In sample C the most abundant compound is thymol (43.5%) while estragole (methylchavicol) predominates in sample T with a content of 35.72%. Eugenol is in the majority in both samples but in different proportions: 8.1% (sample C) and 5.1% (sample T).

Several factors may explain the observed differences in the chemical composition of these two samples, including the time and place of harvesting of the plant, the weather conditions of the vegetative stage, the harvesting method, the drying and storage of the plant material, the part of the plant being studied, the distillation process, the conditions under which the sample is kept, etc. (Burt, 2004; Barbosa *et al.*, 2007).

It is a well-known fact that an essential oil is not reproduced identically in all plants of the same species. It varies according to various elements that condition the life of the plant. These include climatic, histological and ecological elements among others. For example, the same plant species may have two or more chemical races called "chemotypes" or "chemotypes" as described above (Burt, 2004).

The comparison of our results with those of previous work indicates that the essential oil of our sample of *Ocimum canum* corresponds to the chemotype rich in 1,8-cineole; the essential oil of *Ocimum gratissimum* L. corresponds to the chemotype "rich in thymol" while that of our sample of *Ocimum basilicum* belongs to the chemotype "chavicol/linalol".

Although these two oils have already been studied by Cimanga and his collaborators, it is demonstrated that this is the first time that an identification of such a large number of constituents of the latter has been carried out.

After the results of identification of some components of the extracts of three *Ocimum* and their essential oils, the synthesis of some esters was carried out.

3.2. Synthesis of esters

The aim of obtaining the esters is to have the homologues of butyl stearate as lead compound isolated from *Ocimum basilicum and* having a good anti sickle cell disease activity

(Tshilanda et *al.*, 2014) in order to study the structural-reactivity correlation of these different esters.

Four esters were synthesized from palmitic acid in reaction with methanol, ethanol, *n-propanol,* n-butanol according to the operating conditions to obtain: methyl palmitate, ethyl palmitate, n-propyl palmitate, and butyl palmitate .

Two esters were synthesized from oleic acid with methanol and ethanol in turn to obtain: methyl oleate and ethyl oleate.

Four ester mixtures were obtained by reacting technical stearic acid (which is a mixture of stearic and palmitic acids) with methanol, ethanol, *n-propanol* and butanol, these are : mixture 1: methyl stearate + methyl palmitate; mixture 2: ethyl stearate + ethyl palmitate; mixture 3: propyl stearate + propyl palmitate; mixture 4: butyl stearate and butyl palmitate.

Two esters were obtained using acetic acid with :

1) *n-hexanol*: hexyl acetate
2) *n-heptanol*: heptyl acetate

Most of the hexyl formate was obtained by reacting formic acid with *n-hexanol*.

GIC and GIC/MS were used to analyze the synthetic products.

The general mechanism of esterification was the only route of synthesis for all the compounds obtained.

$$R{-}COOH + R{-}OH \underset{}{\overset{H_2SO_4}{\rightleftharpoons}} R{-}COO{-}R + H_2O + H_2SO_4$$

3.2.1. Synthesis from fatty acids

Fatty acids were taken in default and alcohols in excess with 10 drops of sulphuric acid as a catalyst. Ester formation was achieved by carrying out the esterification reaction as shown in the reaction equation above.

3.2.1.1. *From palmitic acid*

(a) synthesis of methyl palmitate

Methyl palmitate was synthesized according to the following equation:

$$CH3\text{-}(CH2)_{14\text{-}COOH} + CH3OH \xrightleftharpoons{H2SO4} CH3\text{-}(CH2)_{14\text{-}COOCH3} + H2O$$

2,059 g of palmitic acid are dissolved in 25 ml of excess methanol (as solvent and reagent); after heating under reflux for 2 h this resulted in 1,818 g of the synthesis product in the liquid (clear) state at room temperature which solidifies at low temperature. Knowing that the molecular weight of the expected product (methyl palmitate) is 270.46 g.mol-1, the calculated yield of the reaction is 83.69%. The molecular formula of methyl palmitate is C17 H34O2.

The good yield of methyl palmitate obtained by this synthesis is due to the quality of the palmitic acid in our possession as well as the ease of removing methanol by washing the reaction medium with water.

(b) Synthesis of ethyl palmitate

Ethyl palmitate was synthesized according to the following equation:

$$CH3\text{-}(CH2)_{14\text{-}COOH+CH3\text{-}CH2OH} \xrightarrow{H2SO4} CH3\text{-}(CH2)_{14\text{-}COOCH2CH3} + H2O$$

The mixture of 2.011g palmitic acid and 30 mL excess ethanol (as solvent and reagent) in a 100 mL flask is heated under magnetic stirring for about 2 reflux. After removal of the excess reagent and catalyst and drying with anhydrous magnesium sulfate, 1.96 g of a clear liquid product at room temperature which solidifies at low temperature has been obtained. Since the molecular weight of the synthesis product is 284.49 g.mol-1, the calculated yield for this reaction is 86.68%. The molecular formula of the product obtained being C18 H36O2.

The good performance of this reaction is also due to the ease with which the ethanol is removed from the reaction medium.

(c) synthesis of *n-propyl* palmitate

N-propyl palmitate was obtained by the following reaction:

$$CH3\text{-}(CH2)_{14\text{-}COOH} + CH3\text{-} CH2\text{-}CH2OH \xrightleftharpoons{H2SO4} 3\text{-}(CH2)_{14COOCH2\text{-}CH2\text{-}CH3} + H2O$$

The mixture of 2.0052g palmitic acid, 35mL *n-propane* and 10 drops sulphuric acid is placed in a 100 mL flask and subjected to reflux heating with magnetic stirring for 2 h. After purification of the reaction medium, 1.99 g of the synthesis product in the liquid state is obtained. The yield of the reaction is 85.55% knowing that the molecular weight of propyl palmitate is 298.52 g.mol-1 and its gross formula is C19 H38O.

The purification of the synthetic product (*n-propyl* palmitate) was made possible by the solubility of the excess reagent (*n-propanol*) in water and by the ease of drying it with magnesium sulphate since the product is in a liquid state.

d) Synthesis of butyl palmitate

N-butyl palmitate was obtained by the reaction below:

$$CH3\text{-}(CH2)_{14\text{-}COOH} + CH3\text{-}CH2\text{-}CH2\text{-}CH2OH \xrightarrow{H2SO4} (CH2)_{14COOCH2\text{-}CH2\text{-}CH2\text{-}CH3} + H2O$$

2.0293g of palmitic acid are put together with 30mL of *n-butanol* and 10 drops of concentrated sulphuric acid in a 100ml flask and subjected to reflux heating for 2.5 hours and magnetic stirring. The product received after all purification steps is clear and liquid, its mass is 1.89g. The yield of the reaction is 76.61% knowing that the raw formula of the synthesized compound is C20 H40O2 and its molecular weight is 312.54 g.mol-.

The butanol taken in this excess reaction is removed with relative ease, due to its partial solubility in water.

3.2.1.2. From stearic acid

(a) Ethyl stearate synthesis (DS2)

The synthesis of ethyl stearate was carried out according to the reaction below:

$$CH3\text{-}(CH2)_{16\text{-}COOH} + CH3\text{-}CH2OH \xrightarrow{H2SO4} CH3\text{-}(CH2)_{16COOCH2\text{-}CH3} + H2O$$

The mixture of 2,5 g of stearic acid, 30 ml of ethanol and 10 drops of concentrated sulphuric acid is placed in a 100 ml flask and heated under reflux with magnetic stirring for 2 hours. The product obtained (oily and clear) has a mass of 2.14 g, which gives a yield of 77.82%, given that the molecular weight of ethyl stearate is 312.54 g.mol-1 and its molecular formula: C20 H40O2.

(b) Synthesis of *n-propyl* stearate (DS3)

The synthesis of *n-propyl* stearate was carried out following the reaction equation below:

$$CH3\text{-}(CH2)_{16\text{-}COOH} + CH3\text{-}CH2CH2OH \xrightarrow{H2SO4} CH3\text{-}(CH2)_{16COOCH2\text{-}CH2CH3} + H2O$$

The mixture: 2g of stearic acid, 25 ml of *n-propanol* and 10 drops of concentrated sulphuric acid was brought to the boil under reflux heating and magnetic stirring for 2 hours. The clear oily product obtained after purification weighed 1.01g, giving a yield of 43.61%. The molecular weight of *n-propyl* stearate being 326.57 g.mol-1 and its gross formula C21 H42O2.

The low efficiency of this reaction would be due to the reduced heating time.

c) Synthesis of isopropyl stearate (DS3')

The synthesis of isopropyl stearate was obtained according to the following reaction equation:

$$CH3\text{-}(CH2)_{16\text{-}COOH} + (CH3)_{2\text{-}CH2OH}\ CH3\text{-} \underset{}{\overset{H2SO4}{\longleftarrow}} (CH2)_{16COOCH2}\text{-}(C\ H3)_2 + H2O$$

And following the procedure below:

The mixture of 4.0803g of stearic acid, 35 mL of isopropanol and 10 drops of concentrated sulphuric acid is placed in a 100 mL flask and subjected to reflux heating with magnetic stirring for 2 h. The product obtained weighs 2.85 g and the calculated yield is 65.75%. The molecular weight of isopropyl stearate and the crude formula being the same as that of *n-propyl* stearate,

The product of this reaction (isopropyl stearate) is pasty like fat, unlike *n-propyl* stearate which is oily,

d) Synthesis of *n-butyl* stearate (DS4)

Butyl stearate was obtained by following the reaction equation below:

$$CH3\text{-}(CH2)_{16\text{-}COOH} + CH3\text{-}(CH2)_{2CH2OH}\ CH3\text{-} \underset{}{\overset{H2SO4}{\rightleftarrows}} (CH2)_{16COOCH2}\text{-}(CH2)_{2CH3} + H2O$$

And the following M.O:

2.50g of stearic acid are dissolved in 30 mL of *n-butanol* and 10 drops of concentrated sulphuric acid are added to the mixture; the mixture is heated with reflux and magnetic stirring in a 100mL distillation flask for 2 h. The end of the reaction is observed by the absence of water drops in the Dean-Stark apparatus.

The liquid, light brown synthetic product has a mass of 1.70 g. Knowing that the molecular weight of butyl stearate is 340.60 g.mol-1 and its molecular formula C22 H44O2, the yield of the reaction was 54.73%.

For esters prepared from two saturated fatty acids, the shorter the alcohol chain, the more solid the ester formed in the absence of light.

3.2.1.3. Synthesis from oleic acid

a) Synthesis of Methyl Oleate (DL1)

Methyl oleate is obtained by following the reaction equation below:

$$CH3\text{-}(CH2)_{7\text{-}CH=CH}(CH2)_{7COOH} + CH3OH \xrightarrow{H2SO4} CH3\text{-}(CH2)_{7\text{-}CH=CH}(CH2)_{7COOCH3} + H2O$$

according to the following procedure:

The mixture of 5ml of oleic acid corresponding to 4.47g of oleic acid, 20ml of methanol and 10 drops of concentrated sulphuric acid is placed in a 100mL distillation flask and then subjected to reflux heating with magnetic stirring for 2 hours. The product of the reaction is a yellow liquid with a mass of 3.81g corresponding to a good yield of 80.85%, knowing that the molecular weight of methyl oleate is 297.47 g.mol-1 and its gross formula is C19 H36O2.

b) Synthesis of DL3 ethyl oleate

The synthesis of ethyl oleate was carried out using the following reaction equation:

$$CH3\text{-}(CH2)_{7\text{-}CH=CH}(CH2)_{7} COOH + CH3\text{-}CH2OH \xrightarrow{H2SO4} CH3\text{-}(CH2)_{7\text{-}CH=CH}(CH2)_{7CO2CH2CH3} + H2O$$

and the following procedure:

The mixture of 5ml of oleic acid corresponding to 4.47grs of oleic acid, 20mL of methanol and 10 drops of concentrated sulphuric acid is placed in a 100mL distillation flask and then subjected to reflux heating under magnetic stirring for 2 hours. The synthesis product obtained after all the purification steps is an orange liquid whose mass is 4.89g corresponding to a yield of 94.75%, the molecular weight of ethyl oleate and its gross formula being C20 H38O2.

The yield is high for ethyl oleate, indicating that all operating conditions were favourable for this reaction.

c) Synthesis of n-propyl oleate

The n-propyl oleate was obtained by following the reaction equation below:

$$CH_3\text{-}(CH_2)_{7}\text{-}CH=CH\ (CH_2)_{7COOH} + CH_3\text{-}CH_2CH_2OH \xrightleftharpoons{H2SO4}$$
$$CH_3\text{-}(CH_2)_{7}\text{-}CH=CH\ (CH_2)_{7CO}\ OCH_2CH_2CH_3 + H_2O$$

Thus, 5ml of oleic acid corresponding to 4.47g of oleic acid, 25mL of n-propanol and 10 drops of concentrated sulphuric acid were placed in a 100 mL flask and subjected to reflux heating accompanied by magnetic stirring for 2 hours.

The reaction product received after purification of the reaction medium is a brown liquid tending to red with a mass of 3.18g. The molecular weight of *n-propyl* oleate being 340.56 g.mol-1 and its gross formula C21 H40O2.

The yield of the reaction is 59.04%.

The esters obtained from oleic acid are all liquid.

3.2.2. Synthesis from simple acids

Simple acids (formic acid and acetic acid) were taken in excess while heavy alcohols (*n-hexanol* and *n-heptanol)* were taken in deficiency.

3.2.2.1. Synthesis of hexyl formate

The synthesis of hexyl formate was made possible by following the reaction equation below:

HCOOH + CH3-CH2CH2CH2CH2OH HCOOCH2 (CH2)₄ CH3 +H2O

Hence, a mixture of 15ml formic acid, 10ml *n-hexanol* and 10 drops of sulphuric acid is placed in a 100ml flask and subjected to reflux heating with magnetic stirring for 2 hours.

The clear liquid product obtained weighs 9.40 mg and the reaction yield of 90.53% would indicate a great success of this reaction. The molecular weight of hexyl formate is 130.19 g.mol-1 and its gross formula is C7H14O2.

The *n-hexyl* formate formed is accompanied by traces of unreacted *n-hexanol-1*, which is difficult to remove by washing with water.

3.2.2.2. Synthesis of hexyl acetate (DA6)

The synthesis of hexyl acetate was carried out according to the following reaction equation

afterwards:

$$CH3COOH + CH3\text{-}CH2CH2CH2CH2OH \underset{}{\overset{H2SO4}{\rightleftarrows}} CH3COOCH2\ (CH2)_{4CH3} + H2O$$

Thus, 15ml of *n-hexanol*, 15ml of acetic acid and 10 drops of sulphuric acid are mixed in a 100 ml flask, which is subjected to reflux heating for 3 h under magnetic stirring.

The product obtained has a mass of 15.08 g. Knowing that the molecular weight of hexyl acetate is 158.24 g.mol-1 for a crude formula C9 H18 O2, the yield of the reaction is 87.58%.

As before, the hexyl acetate formed would be accompanied by *n-hexanol*, a heavy alcohol which would not have reacted and was not separated from the product formed.

3.2.2.3. Synthesis of heptyl acetate (DA7)

The n-heptyl acetate was obtained by following the reaction equation below:

$$CH3COOH + CH3\text{-}(CH2)_{5\text{-}CH2OH}\ \overset{H2SO4}{\longleftarrow} CH3CO\cancel{OCH2\ (}CH2)_{5CH3} + H2O$$

Thus, 15ml of *n-heptanol*, 15ml of acetic acid and 10 drops of sulphuric acid are mixed in a 100 mL flask which has been subjected to reflux heating for 3 h with magnetic stirring. The product obtained (clear liquid) weighs 15.06 g and the calculated yield is 89% knowing that the molecular weight of n-heptyl acetate is 158.24 g.mol-1 for a crude formula C9 H18O2.

The synthetic *n-heptyl* acetate would be accompanied by traces of unreacted *n-heptanol, which* is very poorly soluble in water.

3.2.3. Analysis of esters obtained by GC and GC/MS

3.2.3.1. Analysis of esters from simple acids

(a) Analysis of esters from formic acid and acetic acid

1) Hexyl form (DF6)

On the chromatogram of the DF6 sample, which should be the hexyl formate (Figure 3.65), a large, broad peak is visible and ranges from 6 to 6.4 min as retention time; other small peaks are visible at RT= 8.134min, 7.455 min.

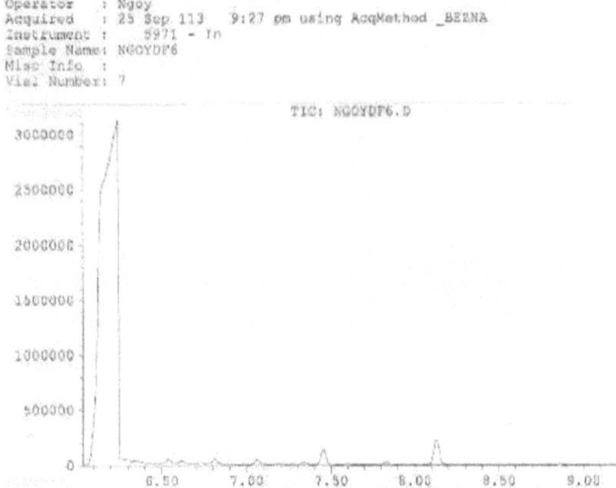

Figure 3.65. DF6 chromatogram

The large peak between 6 and 6.2 min is a mixture of two products, one (TR=6.150 min) is *n-hexanol and the* other (TR=6.184 min) is hexyl formate, which is the expected product of this reaction; the small peak at TR=8.136 min is octyl formate based on a comparison of their mass spectra received by GC/MS and those of their counterparts in the Wiley 275 database. The mass spectrum of DF6 hexyl formate compared to the database is shown in Figure 4.23 in the Appendix.

Other small peaks can be considered as impurities.

The resulting hexyl formate is contaminated with *n-hexanol,* which has not been dissolved in water for removal from the reaction medium (Figure 4.24 in the Appendix).

2) Hexyl acetate (DA6)

For the DA6 sample which should be hexyl acetate; 7 peaks are visible on its chromatogram of the sample attesting the presence of 7 compounds in this sample, and of which only one has a very high intensity at retention time TR = 6.921min   testifying that it is the only major component (**Figure 3.66**).

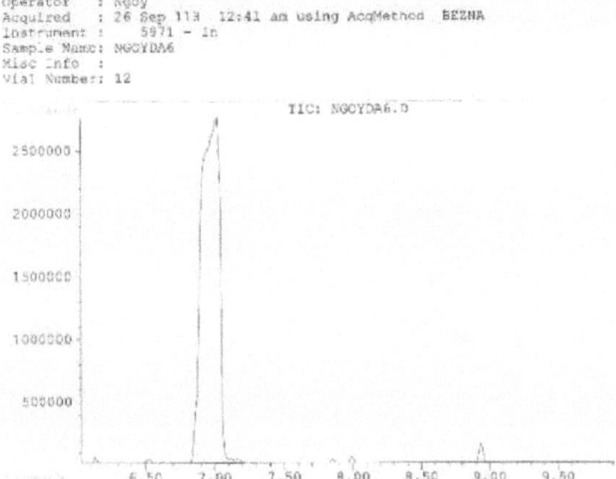

Figure 3.66: Chromatogram of DA6

Gas chromatography coupled to the sample mass spectrometer and comparison with the mass spectra in the database indicates that the majority compound is hexyl acetate, which is likely to be contaminated with hexanol given the width of the peak. The mass spectrum of this compound is shown in the appendices (Figure 4.25.).

3) *n-heptyl* acetate (DA7)

For sample DA7 which would be heptyl acetate; 3 peaks are visible on its chromatogram (Figure 3.67) attesting the presence of 3 compounds in this sample, and of which only one (TR=8.142min) has a very high intensity attesting that it is the only major component.

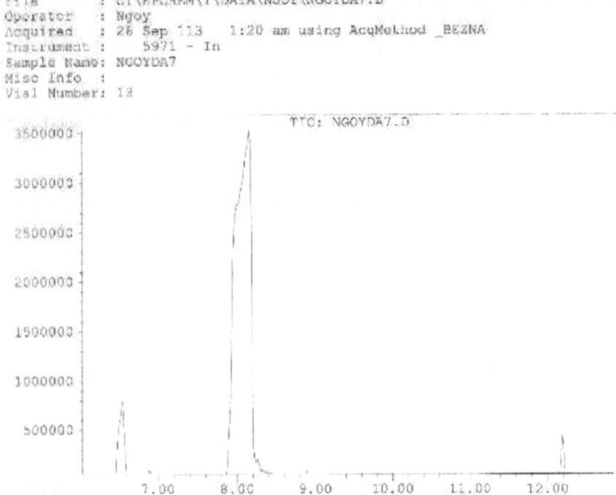

Figure 3.67. : Chromatogram of DA7

GC/MS coupling of the sample and its comparison to the mass spectra in the database indicates that the major compound is heptyl acetate.

The mass spectrum of the majority component compared to its counterpart in the database is in the appendices (in Figure 4.26).

3.2.3.2. Analysis of esters from palmitic acid

(a) Methyl palmitate (DP1)

A single peak at RT=15,508 min is observed on the GC chromatogram (Figure 3.68) **of the DP1** sample which should be methyl palmitate, indicating that the reaction resulted in a single product whose GC/MS mass spectrum compared to its counterpart in the database actually identifies it with methyl palmitate.

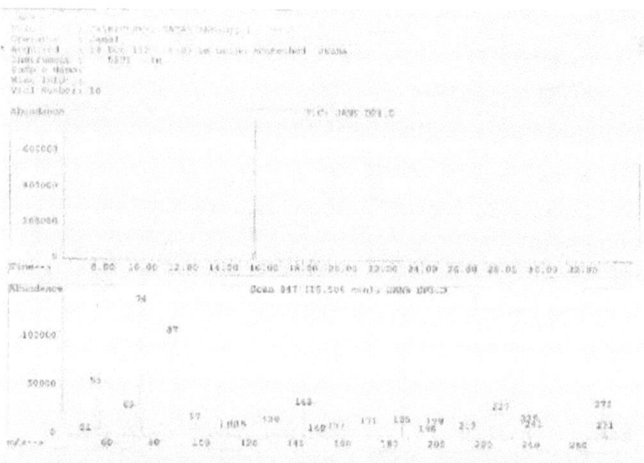

Figure 3.68: DP1 chromatogram and compound mass spectrum

From the above, it is clear that the palmitic acid in our possession is pure and the operating conditions have favoured the synthesis of methyl palmitate.

(b) Ethyl palmitate (DP2)

For the sample that should contain ethyl palmitate, a single peak is visible at TR= 16.321 min on the chromatogram of the DP2 sample showing the presence of a single compound in this sample (Figure 3.69).

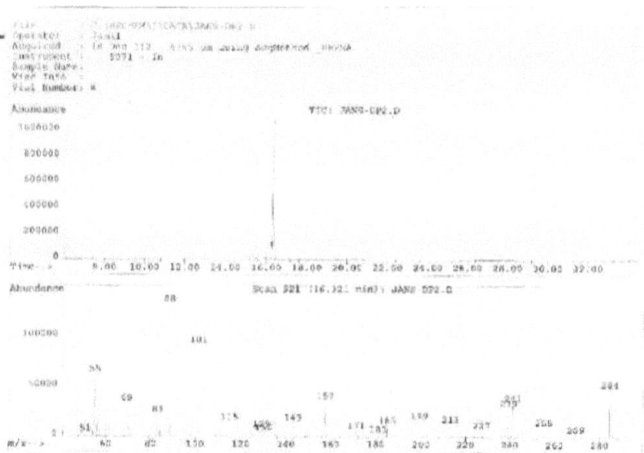

Figure 3.69. DP2 Chromatogram and Product Mass Spectrum

The mass spectrum of this peak obtained by gas chromatography coupled to the mass spectrometer and its comparison with its counterpart in the database indicates that the compound obtained is ethyl palmitate.

The operating conditions and the good quality of palmitic acid allowed the synthesis of ethyl palmitate.

(c) *n-Propyl* palmitate (DP3)

For the sample that should be *n-propyl* palmitate, a single peak was visible (TR=17,250 min) on the chromatogram of the DP3 sample confirming the presence of a single product obtained after this reaction. Figure 3.70 shows the mass spectrum of the DP3 product and that of the database.

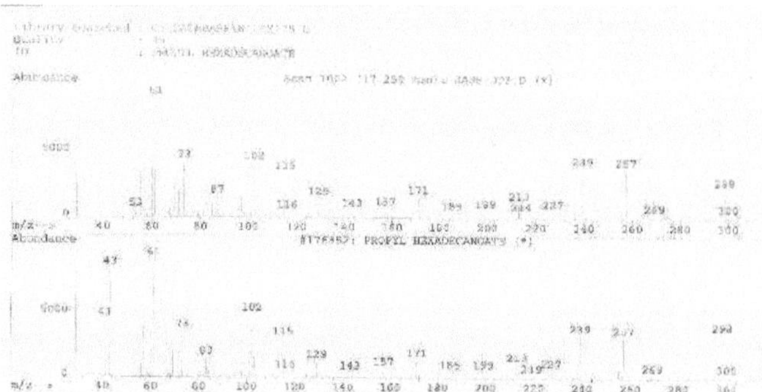

Figure 3.70. : Mass spectrum of *n-propyl* palmitate compared to DP3

Comparison of these two mass spectra allows the identification of the *n-propyl* palmitate reaction product, the database being Wiley 275 L. It is therefore clear that the synthesis has arrived at the expected product.

(d) *n-Butyl* palmitate (DP5)

For the sample assumed to contain *n-butyl* palmitate; a single peak (TR=18,916 min) is visible on the chromatogram of the DP5 sample showing the presence of a single compound in this sample (Figure 3.71).

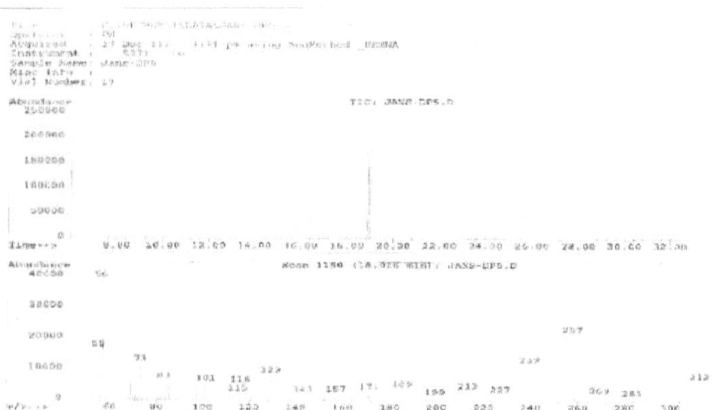

Figure 3.71. DP5 chromatogram and mass spectrum of the majority compound

Gas chromatography coupled to the sample mass spectrometer and comparison with the mass spectra in the database indicates that the majority compound is *n-butyl* palmitate. Thus, *n-butyl* palmitate was successfully obtained by esterification.

From these four reactions with palmitic acid, it can be seen that the more the chain of the product obtained increases, the more the retention time TR increases. The latter increases by about one minute by adding a carbon in the order: 15,508, 16,321, 17,250 and 18,916 min.

3.2.3.3. Analysis of stearic acid esters

(a) Ethyl stearate (DS2)

Two large peaks are visible on the chromatogram of the sample which should be ethyl stearate confirming the presence of two major compounds, the most important of which is the one at RT= 15.97 min (given the peak intensity) followed by the one at RT= 19.08 min.

The presence of minority peaks is also observed at TR= 13.88 min, at about 14.80 min, 17.30 min, between 18 and 19 min, at about 21 min and 24 min (Figure 3.72).

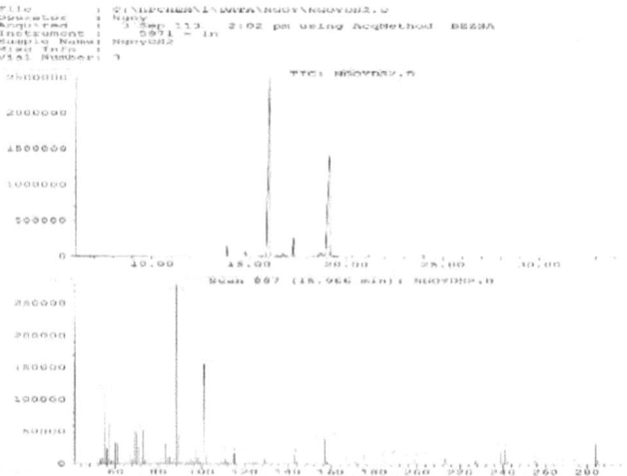

Figure 3.72. DS2 chromatogram and the mass spectrum of the product at RT=15,966 min.

Gas chromatography coupled to the mass spectrometer of the synthetic sample and its comparison with the mass spectra in the database indicates that the two majority compounds are: ethyl palmitate (RT=15.97 min), followed by ethyl stearate (RT= 19.08 min). Two small peaks were identified with ethyl heptadecanoate (TR=17.30 min) and ethyl myristate (TR=13.88 min).

All compounds identified in **DS2** are esters, indicating that the impurities in stearic acid are other fatty acids and that they all reacted in the presence of ethanol.

(b) *n-Propyl* stearate (DS3)

The major peaks visible on the gas chromatogram of the **DS3** sample to be *n-propyl* stearate prove that the synthesis resulted in a mixture of at least two products in the following descending order (by comparing the intensity of their peaks on the gas chromatogram): that at TR=17.39 min, 21.46 min, 15.71 min, at about 20.10 min, 14.90 min, 18.6 min, 23.9 min, 27.6 min (Figure 3.73).

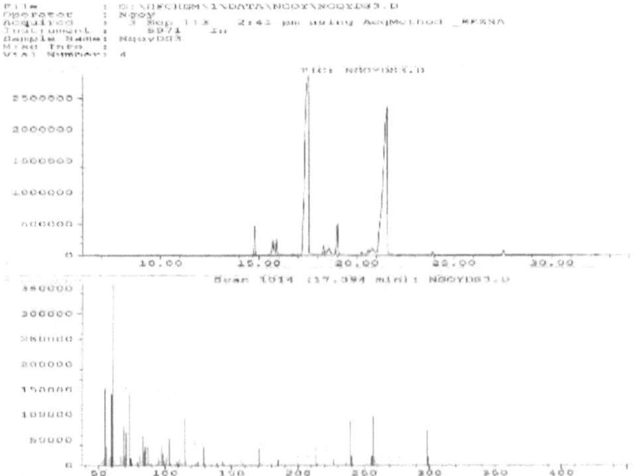

Figure 3.73: Chromatogram of DS3 and the mass spectrum of the compound at RT=17,394 min

The mass spectra of a few compounds in the sample obtained by gas chromatography coupled to the mass spectrometer are compared with those in the database indicating that the majority compounds are: *n-propyl* palmitate (17.39 min) followed by *n-propyl* stearate (21.46 min), two small peaks are identified with palmitic acid (TR= 15.71 min) and heptadecanoic or margarinic acid (TR=18.58 min). The mass spectra of two majority compounds are shown in the appendices (Figure 4.28).

The content of *n-propyl* stearate is close to that of *n-propyl* palmitate and appears to be higher in **DS3** than ethyl stearate in **DS2**. The impurities in **DS3** are fatty acids that have not undergone esterification and are trace amounts.

(c) Butyl stearate (DS4)

Two large peaks are visible on the gas chromatogram of the **DS4** sample, the largest of which is at TR=18.56 min followed by the one at TR=23.76 min. Small peaks are observed around TR= 21.0, 15.8, 9.2 and 16.2 min, indicating that the reaction resulted in a mixture of products (Figure 3.74).

```
File        : C:\HPCHEM\1\DATA\NGOYDS4.D
Operator    : Ngoy
Acquired    : 13 Aug 113   10:58 pm using AcqMethod _BEZNA
Instrument  :    5971 - In
Sample Name: NGOYDS4
Misc Info   :
Vial Number: 15
```

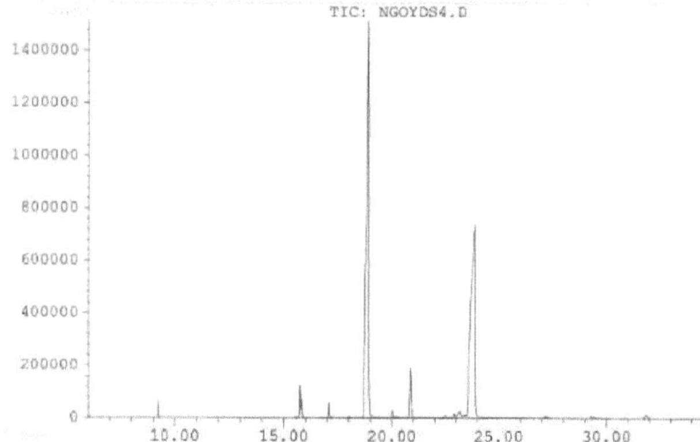

Figure 3.74. DS4 chromatogram

By GC/MS and comparison of the mass spectra of the components of DS4 with those in the database, the single peak at TR=23,762 min was identified as butyl stearate. The strongest peak should be butyl palmitate in comparison with the other **DS** samples. The mass spectrum of butyl stearate from DS4 compared to the database is shown in Figure 4.29 in the Appendix.

Two main compounds were obtained during this reaction (taking into account the intensities of their peaks) and the least abundant is the expected product during this reaction (butyl stearate); the most abundant should be butyl palmitate (mass 312 amu) but which has not been identified.

d) Isopropyl stearate (DS6)

Two large peaks on the gas chromatogram of DS6 (TR=16.21 min and 19.45 min) prove that the reaction led to two major products instead of one (Figure 3.75).

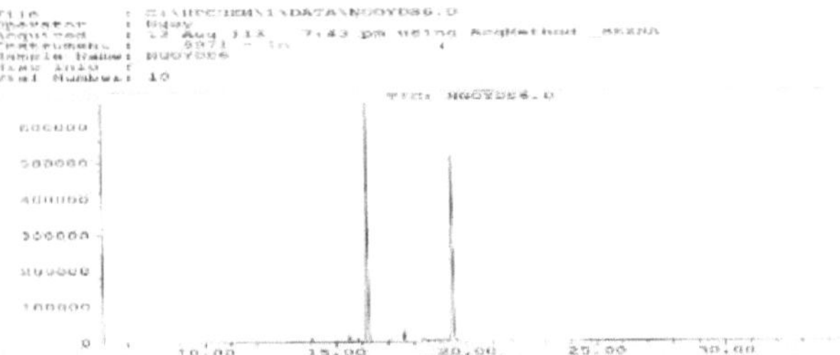

Figure 3.75. DS6 chromatogram

These two compounds are identified with isopropyl palmitate (RT=16.21 min) and isopropyl stearate (RT=19.45 min)   by comparing their mass spectra obtained by GC/MS with those in the database.

The almost exclusive presence of the esters of two fatty acids contained in what should be the stearic acid in possession, shows that the reaction was complete and the very small peaks being considered as impurities.

3.2.3.4. Analysis of esters from oleic acid

(a) Methyl oleate (DL1)

The presence of three peaks visible on the gas chromatogram of DL1, the largest of which is at TR=17.46 min and two small peaks at TR=14.95 min and 15.14 min, indicates that the reaction led to a major product contaminated by two small impurities (Figure 3.76).

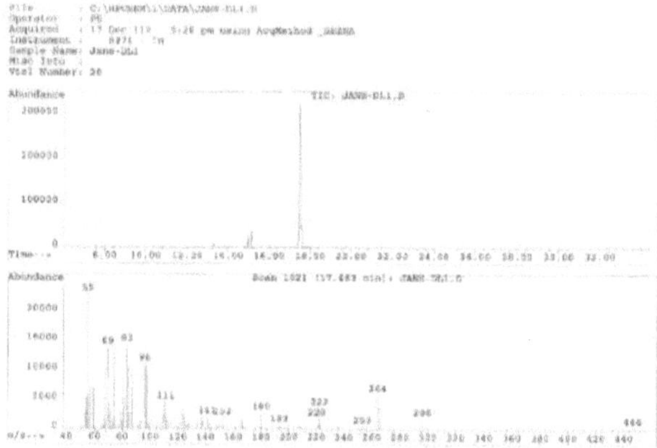

Figure 3.76. Chromatogram of LD1 and the mass spectrum of the compound at RT=17.46 min.

Comparison of the mass spectrum of the majority compound with that of its counterpart in the Wiley 275L database identifies it as 9-methyl oleate or methyl otadecenoate, which is the expected product for this reaction.

(b) Ethyl oleate (DL3)

A large peak (RT=20.58 min) is accompanied by three small peaks (RT=16.9, 17.0, 20.7 min) on the gas chromatogram showing that the majority product obtained during this synthesis is accompanied by three small products considered as impurities (Figure 3.77).

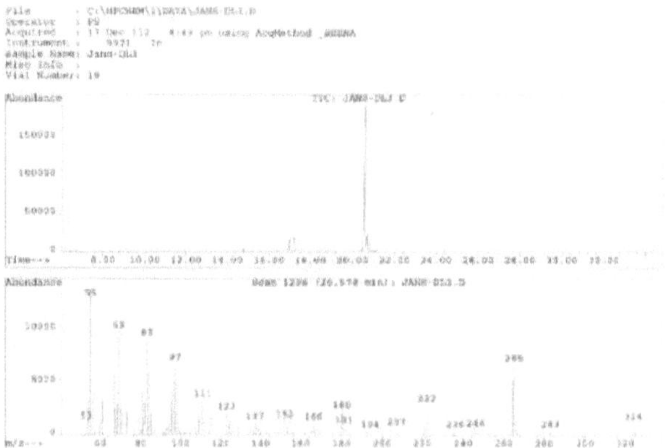

Figure 3.77. : Chromatogram of LD3 and the mass spectrum of the major constituent (TR= 20.578 min)

The major compound is identified by comparison of its GC/MS mass spectrum with its counterpart in the Wiley275L database at (Z) 9-ethyl oleate or ethyl otadecenoate which is the expected compound for this synthesis.

The success of this synthesis is noted by the fact that the major product present is the one expected and that the small impurities would be those in the oil in our possession.

It is noted that after this synthesis, the success was for the esters obtained from palmitic and oleic acid while those obtained from stearic acid were in a binary mixture where the palmitic ester dominated, showing that stearic acid was in a mixture with palmitic acid. Similarly, due to the difficulty of removing heavy alcohols (*n-hexanol* and *n-heptanol*) from the reaction medium, although they were found to be deficient, the esters obtained were accompanied by the alcohol, which is why the formic and acetic acid esters were not used for the sickle cell disease test in the continuation of this work.

Given the presence of some methyl esters in our reserve, they were considered together with those obtained synthetically to evaluate their anti sickle cell activity. Their list can be found in the appendix (Table 4.2).

3.3. Biological activities

3.3.1. Antifalcemic activity

3.3.1.1. Antifalcemic activity of plants

a) Antifalcemic activity of methanolic extracts

Figure **3.78** below gives the phenotype of SS erythrocytes alone (a) or treated with methanolic extracts of *O. basilicum* (b), *O. canum* (c) and *O. gratissimum* (d) under hypoxic conditions.

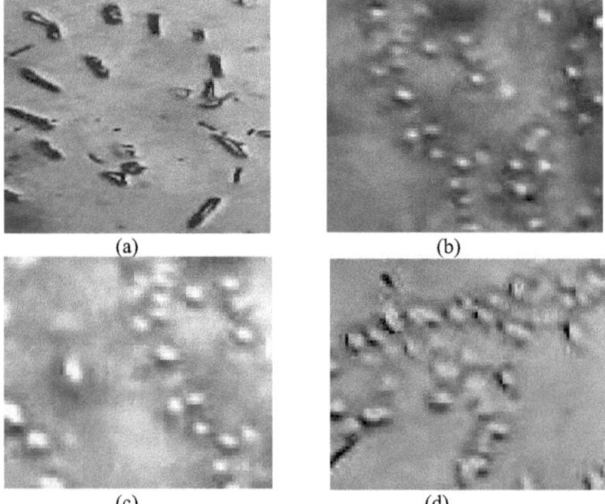

(a)   (b)

(c)   (d)

Figure 3.78: Microphotography of SS blood (a: control) and SS blood incubated with methanolic extracts (b: *Ocimum basilicum*; c: O. *canum;* d: O. *gratissimum*). (x500) [NaCl 0.9%; Na2S2O5 2%].

Figure 3.78a shows that this is indeed SS blood, as many sickle cell erythrocytes can be observed. On the other hand, figures 3.78b, 3.78c and 3.78d indicate that in the presence of methanolic extracts of the three *Ocimum the* majority of erythrocytes are normalized. This shows that methanolic extracts have an antifalcemic activity.

Since rosmarinic acid was identified in each of the methanolic extracts of these three *Ocimums*, its anti-sickle cell activity was also evaluated. Figure 3.79 below shows SS blood treated with a solution of pure rosmarinic acid (AR).

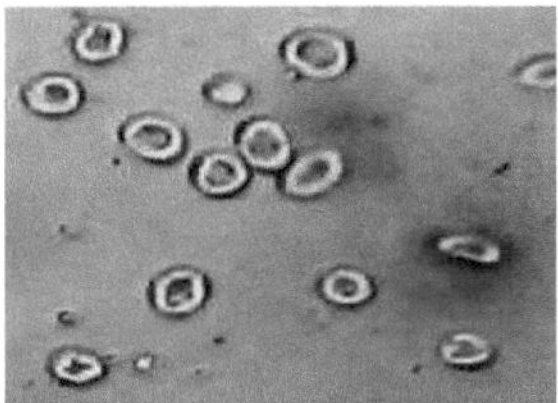

Figure 3.79: Morphology of sickle cells treated with pure rosmarinic acid solution (X500) (20µg/mL) [NaCl 0.9%; Na2S2O4 2%].

It appears from this figure that rosmarinic acid is biologically active and would be the active ingredient in all three species (Tshilanda et *al.,* 2016b).

Similar results were also obtained for other plant extracts used to treat sickle cell disease in traditional Congolese medicine, thus justifying their traditional use (Shode *et al.,* 2011; Mpiana *et al.,* 2014; Tshilanda et al., 2014; Tshilanda *et al.,* 2015a,c).

Figure 3.80 shows the evolution of the normalization rate of SS erythrocytes as a function of the concentration of active extract and rosmarinic acid.

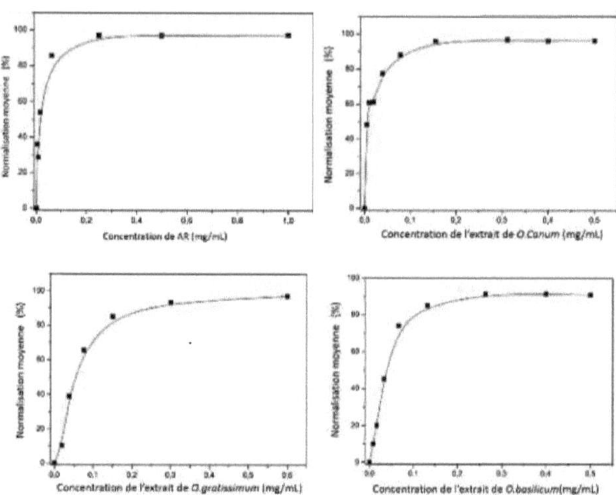

Figure 3.80. Evolution of the normalization rate of sickle cell normalization in the presence of methanolic extracts of *Ocimum* and rosmarinic acid (RA).

Figure 3.80 shows that the normalization rate of sickle cell disease is dose-dependent and follows a hyperbolic pattern.

The minimum normalization concentrations i.e. the lowest concentration of the extract or RA for which the normalization rate is maximum are grouped in the following table.3.34.

Table 3.34: Minimum Normalizing Concentration (MNC) of different plants and rosmarinic acid

Sample	MNC (mg/mL)
Rosmarinic acid	0,18±0.03
Ocimum basilicum	0,23±0.04
Ocimum gratissimum	0,26±0.04
Ocimum canum	0,31±0.05

As noted in Table 3.34, rosmarinic acid (RA) has a greater antifalcemic effect than methanolic extracts. This indicates that the main component responsible for the activity in extracts would be rosmarinic acid (Tshilanda *et al.*, 2016b). Indeed, it can be noted that *Ocimum basilicum*, which has a high RA content, exhibits greater antiseptic activity followed by *O. gratissimum* and *O. canum*.

It is therefore for the first time that such a comparative study of anti sickle cell activity between these three plants has been carried out and it is also the first time that the same activity has been proven for rosmarinic acid (Tshilanda et *al.*, 2016b).

In fact, the first molecules isolated from plants with anti-prepanocyte properties were organic acids (Gorecki et al., 1980; Russu *et al.*, 1986; Akoje & Fung, 1992; Tshibangu, 2011; Tshilanda et al., 2015b; Tshilanda et al., 2016b).

The already known antioxidant activity of rosmarinic acid (Rerhow *et al.*, 2012; Costa *et al.*, 2012) should help to relieve sickle cell disease patients because sickle cell disease is known to generate oxidative stress (Sess *et al.*, 1992; Monnet *et al.*, 2002).

b) Antifalcemic activity of the extracts obtained with 4 solvents with increasing polarity of 3 species

In order to check whether the antifalcemic activity depends on the polarity of the secondary metabolites, we carried out a fractional extraction using solvents with increasing polarity.

Table 3.35 below shows the results of the Sickle Cell Red Blood Cell Inhibition Test as a function of the polarity of the extraction solvents.

Table 3.35. Sickle cell erythrocyte inhibition test result by polarity of secondary metabolites

Type of extracts	Inhibition of adulteration		
	Ocimum basilicum	*Ocimum canum*	*Ocimum gratissimum*
n-hexane	–	–	±
Dichloromethane	–	–	+
Ethyl acetate	++	++	++
Methanol	++	++	++

Legend : (+/-) normalization <10%; (+) **10<normalization** <50%;
(++) normalization >70%

From Table 3.35, it appears that the polar extracts of three plants showed a high activity against sickle cell disease and only the apolar extracts of *Ocimum gratissimum* showed low and medium activity.

The anti sickle cell activity of apolar extracts observed in this study would be due to triterpenoic acids (Tshilanda et *al.,* 2015b; Tshibangu, 2011).

However, Mpiana and colleagues have shown that this activity is also due to anthocyanins (Mpiana et *al.,* 2007; Mpiana *et al.,* 2008). These polar compounds are generally more concentrated in polar solvents such as methanol and ethyl acetate and are believed to be acylated with organic acids in the plant cell vacuole (Ngbolua, 2012).

c) Antifalcemic activity of aqueous, AcOEt, MeOH and anthocyanin extracts of *O. canum*

Since *O. canum* is one of two species not cited in traditional medicine for the management of sickle cell disease, it was selected using a chemotaxonomic approach for validation and inclusion in the list of anti-sickle cell disease plants. Thus, the antifalcemic test was carried out on its extracts: aqueous, anthocyanic, methanolic and acetic.

Figure 3. 81 gives the phenotype of sickle-cells untreated (a) or treated with *O. canum* extracts (b-e) under hypoxic conditions.

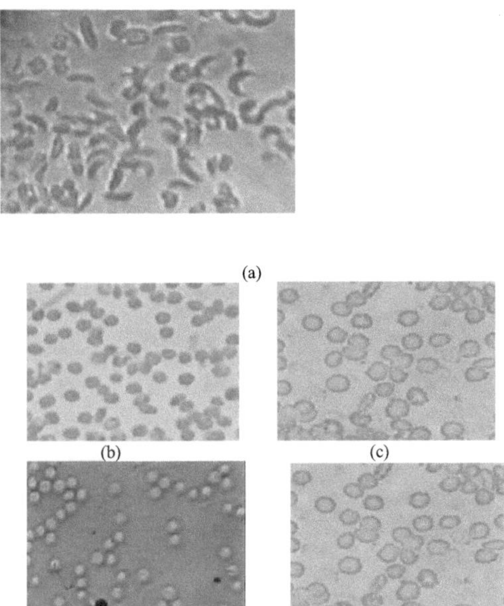

Figure 3.81.: Morphology of red blood cells of SS blood (Control: a) and SS blood incubated with aqueous extracts (b), AcoEt (c), MeOH (d), and anthocyanins (e) of *O. canum*; (X500), [NaCl 0.9%; Na2S2O4 2%].

It can be seen from this figure that the control contains the majority of sickle cells confirming the nature of the sickle cell blood (Fig. 3.81a). When SS blood is brought into contact with the aqueous (Fig. 3.81b), AcoEt (Fig. 3.81c), MeOH (Fig. 3.81d) and anthocyanin extracts (Fig. 3.81e) of *O.canum, the* majority of erythrocytes return to the normal circular shape. This fact indicates that these extracts are endowed with anti sickle cell properties (Tshilanda et *al.*, 2015c).

The antiseptic activity of the anthocyanin extract confirms what has already been observed for some plants used in traditional Congolese medicine (Mpiana *et al.*, 2013; Ngbolua *et al.*, 2015;; Mpiana et *al.*, 2007b). Indeed, anthocyanins interact with haemoglobin S (HbS) and compete with the polymerization reaction and thus inhibit the process of erythrocyte falciformation (Mehanna et al., 2001; Tshilanda et *al.*, 2015c).

Figures 3.82 and 3.83 show the evolution of the normalization of sickle-cells as a function of the concentration of methanolic and anthocyanin extracts of *Ocimum canum* leaves respectively.

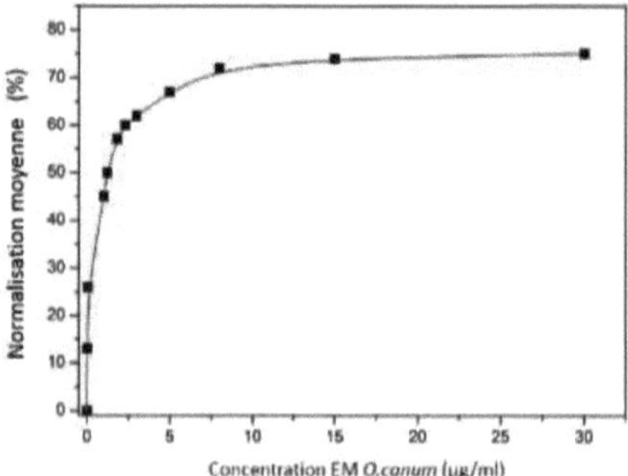

Figure 3.82. Evolution of the rate of   normalization of sickle cell normalization with the concentration of methanolic extract of *O. canum* leaves.

Figure 3.82 shows that the rate of normalization increases with the dose or concentration of the methanolic extract until a maximum threshold is reached at which normalization remains constant even if the concentration of the extract is further increased. The minimum normalization concentration, i.e. the lowest concentration of the extract for a maximum normalization rate of 70%   , is 14 µg/mL (Tshilanda et *al.*, 2015c).

It is therefore noted that the NMC of the total methanolic extract of *O. canum* (Table 3.40) has a higher value (0.31 mg/mL) than that of its methanolic extract from extraction with solvents of increasing polarity. Fractional extraction would improve the antiseizure activity by sequentially removing non-active compounds that would mask the activity.

Figure 3.83 shows the evolution of the normalization rate of sickle cell normalization as a function of the concentration of anthocyanin extracts.

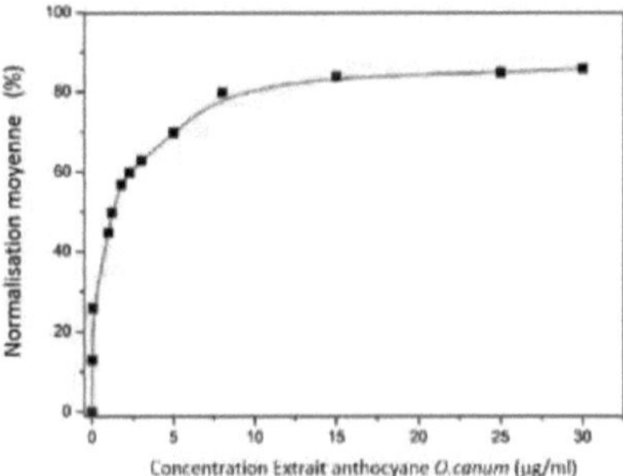

Figure 3.83. Evolution of the normalization rate of sickle cell disease as a function of the concentration of anthocyanin extract.

This figure shows that normalization increases with the concentration of anthocyanin extract and reaches a constant value from 15 µg/mL corresponding to the normalization rate of 88.5%.

These results show that polar extracts show greater activity than non-polar extracts and that this activity is also due to anthocyanins, confirming the earlier work of Mpiana and colleagues (Mpiana *et al.*, 2008; 2013; Ngbolua et *al.*, *2013*).

In another study, Tshibangu(2011) showed that triterpenoic acids also have anti-sickle cell activity. The apolar compounds generally concentrate in dichloromethane. For this purpose, *O. gratissimum* whose apolar extracts (n-hexane, dichloromethane) have shown some activity as well as *O. basilcum* (plant used in traditional medicine) were chosen to test the antifalcemic activity of isolated fractions and products.

The compound Do8 from *O. basilicum* is obtained by precipitation of the extract with methanol acidified with petroleum ether followed by crystallization and recrystallization.

d) Antifalcemic activity of products and fractions obtained from *O. gratissimum* and *O. basilicum*

Table 3.36 shows the results of bio-guided fractionation of two plants.

Table 3.36. Result of bio-guided fractionation of *O. basilicum* and *O. gratissimum*: test for inhibition of sickle cell erythrocyte falciformation by some compounds and fractions from *Ocimum basilicum* and *Ocimum gratissimum*.

Fraction or product	Source	Activity
Ursolic acid	Extract AcOEt *Ocimum gratissimum*	++
Oleanolic acid	Extract DCM *Ocimum gratissimum*	++
EMA	MeOH/AcOEt *Ocimum gratissimum* extract	++
Do8	*Ocimum basilicum*	++
Dina 1	Extract DCM *Ocimum gratissimum*	+
Dina 2	Extract DCM *O.gratissimum*	-
Dina 4	Extract DCM *O.gratissimum*	+
DM1	Extract MeOH *O.gratissimum* (EMA)	++

Legend: -: no normalization; (+/-) normalization <10%; (+) **10<normalization** <50%; (++) normalization >70%

Table 3.36 shows that of four fractions from the DCM extract only oleanolic acid shows good activity, the test on the Dina2 sample is negative and activity below 50% is observed for the other two (Dina1 and Dina 4); whereas all samples from the polar extracts show good activity.

Thus, the antifalcemic activity observed for the DCM extract would be mainly due to the presence of oleanolic acid in this extract, followed by Dina1 and Dina4 of which some of the majority compounds identified are: epi-α selinene, β-bisabolene and trans- caryophyllene for Dina 1 and 2-phenyl-2 typyl acenaptenone for Dina 4.

Similarly, the activity of Do8 would justify that observed for the anthocyanin extract of *Ocimum basilicum* (Tshilanda *et al.*, 2014).

The activity of the EMA extract would be due to the presence of rosmarinic acid in this extract and constituents of DM1 which are : 1,8-dimethoxy-1,8-methyl-3-anthraquinone, triacontane and hexatriacontane and those identified for DM2 and DM3.

170

The antifalcemic activity observed for the ethyl acetate extract of *O. gratissimum* *is* thought to be due to the presence of ursolic acid and this is the first time that the anti sickle cell activity of this compound has been reported (Tshilanda et *al.*, 2015b).

e) *Antifalcemic activity of essential oils and thymol*

Figure 3.84 shows the phenotype of sickle cell disease untreated (a) or treated with the essential oils of three plants (b-d) and with thymol, a major component of the essential oil of *O. gratissimum* (e).

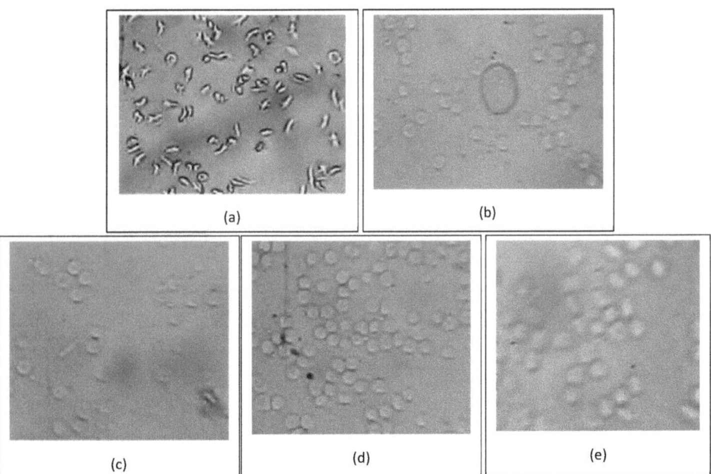

Figure 3.84.: Microphotography of SS blood (a: control) and SS blood incubated with HE from : *O. basilicum* (b): *O. canum* (c), O. *gratissimum*) (d) and pure thymol (e) (x500) [NaCl 0.9%; Na2S2O5 2%].

Figure 3.84a shows that it is actually SS blood with the presence of numerous sickle cell erythrocytes. On the other hand, figures 3.84b, 3.84c, 3.84d and 3.84e indicate that in the presence of the essential oils of the three *Ocimums* and thymol, the majority of erythrocytes return to normal form. This shows that the essential oils and thymol have anti-replacement properties.

This activity would be attributed to methylchavicol (estragole), linalol and eugenol, three major components of the essential oil of *Ocimum basilicum, and* 1,8-cineole would be the basis of that of *O. canum.*

To our knowledge, it is for the first time that the antipanocyte activity of essential oils is reported.

3.3.1.2. Antifalcemic activity of esters

The study of the anti sickle cell disease activity is carried out on the different esters obtained synthetically and those taken in the laboratory (Appendices: Table 4.2) in order to compare their activity with that of butyl stearates isolated from *O. basillcum* (Tshilanda et *al.,* 2014).

Figure 3.85 (a-d) shows respectively the micrographs of SS blood alone in the 0.9% NaCl solution (control: a) and sickle cells treated with a few esters (b-d).

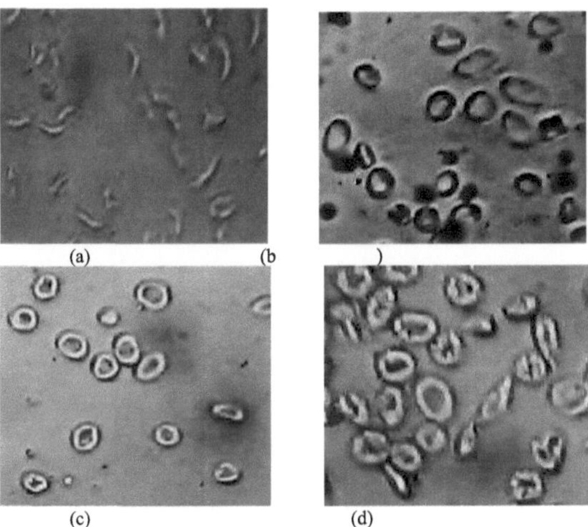

Figure 3.85: Microphotography of SS control blood (a: control) and SS blood treated with some esters (b: butyl palmitate; c: butyl stearate; d: methyl nonadecanoate), (x500) [NaCl 0.9%; Na2S2O5 2%,].

Figure 3.85 shows that control (a) contains predominantly sickle-cells, confirming the nature of SS blood. However, in the presence of the few non-aromatic esters (Fig. 3.85 b-d), the majority of erythrocytes return to normal form. Normalization of sickle cells is much more important when treated with methyl nonadecanoate. This indicates that some synthetic esters have an antifalcemic effect , confirming the result already reported for butyl stearate isolated from *O. basilicum* (Tshilanda et *al.*, 2014).

These results indicate that non-aromatic esters are the best potential candidates for the development of anti-seizure drugs (Tshilanda et *al.*, 2016c).

Knowing that logP, also called log Kow, is a measure of the differential solubility of chemical compounds in two solvents (octanol/water partition coefficient); the theoretical calculation by simulation of this parameter according to the ChemDraw program of a few esters has been carried out and has shown that log P increases with the length of the carbon chain and is a measure of hydrophobicity.

This property allows molecules to pass the biological barrier easily, but the disadvantage is that the more hydrophobic a molecule is, the more it will become trapped in intracellular lipid inclusions. Thus for these different esters, the observed antifalcemic activity would be directly linked to this hydrophobic character and not to an adopted conformation since all molecules are linear.

Table 3.37 below shows the lop p values for each ester considered, their raw formula and their antifalcemic activity.

Table 3.37. Log p values of some esters by chemdraw

Name of the ester	Log value p	Gross Formula	Activity
Methyl undecanoate ·	4,24	C12H24O2	+
Methyl tridecanoate	5,15	C14H28O2	+
Methyl myristate	5,61	C15H30O2	+
Methyl palmitate	6,52	C17H34O2	+
Methyl heptadecanoate	6,98	C18H36O2	+
Methyl stearate	7,43	C19H38O2	+

The estimation of log p using Broto's fragmentation method for methyl myristate in a study by Danny Reible and Katerina Demnerova in 2002 gave the same value (5.61) as that obtained by Chemdraw, thus attesting to the reliability of the Chem Draw program.

Ethyl palmitate has previously been used in combination with snake venom for sickle cell disease (Morag *et al.*, 1970; Oswaldo *et al.*, 1978); however, this is the first time that a study of other fatty acid esters has been conducted (Tshilanda et *al.*, 2016c).

3.3.2. Antibacterial activity of essential oils

EO that have demonstrated antifalcemic activity are also known for their antibacterial properties. Because of functional asplenia, sickle cell patients are generally susceptible to bacterial infections. Compounds with both sickle cell and antibacterial properties would be useful in the formulation of drugs for the better management of sickle cell disease.

Table 3.38 shows the results of the antibacterial test.

Table 3.38: Antibacterial Activity Results

HE and antibiotics	Microbial strains / MIC and MBC in mg/mL																	
	E. coli			*S. aureus*			*K. pneumoniae*			*S. typhimurium*			*Proteus mirabilis*			*Pseudomonas aeruginosa*		
	CMI	WMC	R	CMI	WMC	R	CMI	WMC	R	CMI	WMC	R	CMI	WMC	R	CMI	WMC	R
O. basilicum	0,143	45,9	320,97	1,147	45,9	40,017	1,43438	-	-	0,1793	-	-	11,475	45,9	4,0017	0,1793	45,9	256
O. canum	5,8275	-	-	46,62	-	-	-	-	-	-	-	-	-	-	-	0,182	-	-
O. gratissimum	0,03141	2,01	63,992	0,251125	2,01	8,004	0,5025	4,02	8	0,5025	8,04	16	0,03141	0,125	3,979	0,2515	0,25	1
Gentamicin	0,0039	0,25	64,102	0,0156	-	-	0,125	-	-	0,625	-	-	-	-	-	0,0312	-	-
Cifen	0,00098	0,00781	7,969	0,00196	-	-	0,00196	-		0,062	-	-	0,25	-	-	0,125	-	-
Augmentin	0,00098	-	-	0,00196	-	-	0,25	-		0,00196	-	-	0,062	-	-	0,25	-	-

Legend: MIC: minimum inhibitory concentration

MBC: Minimum bactericidal concentration

A: CMB/CMI ratio; if R is $\leq$ 4: bactericidal effect and if R is > 4: bacteriostatic effect

It appears from this table that *E. coli, S. aureus, K. pneumoniae* and *P. mirabilis* are more sensitive to the essential oil of *O. gratissimum* than to the essential oil of *O. basilicum*. Whereas *S. typhimurium* and *P. aeruginosa are on the* contrary sensitive to the essential oil of *O. basilicum; on the* other hand the essential oil of *O. canum* has only been active on *P. aeruginosa.*

In addition, it should be noted that the essential oil of *O. gratissimum* has shown great activity compared to the standard antibiotics used for *P. mirabilis.* This high antibacterial activity is thought to be due to the presence of thymol.

The inhibition of *Pseudomonas aeruginosa* and *Staphilococus aureus* by *O. gratissimum* essential oil is thought to be due to the presence of thymol and carvacrol in the oil, which in one study inhibited these two bacterial strains (Lambert *et al.,* 2001).

To confirm the nature of the bacterial effects of the oils and antibiotics, MBC/MIC ratios were calculated. Kpodedekon *et al* (2013) reported that this ratio is used to determine the antibiotic potency of the essential oil. When this ratio is less than or equal to 4, the essential oil is said to be bactericidal and when it is greater than 4, the oil is bacteriostatic.

Analysis of the results in Table 3.44 above shows that the essential oil of *Ocimum basilicum* is bactericidal on one strain (*Proteus mirabilis*) and had a bacteriostatic effect on three strains (*E. coli, S. aureus, P. aeruginosa*) while that of *Ocimum gratissimum had* a bactericidal effect on two strains, namely : *Proteus mirabilis* and *Pseudomonas aeruginosa* and a bacteriostatic effect on the other four strains; on the other hand *Ocimum canum showed* no bactericidal effect on all the strains tested.

Of the three antibiotics taken as references none showed a bactericidal effect on all strains; only a bacteriostatic effect is presented against one strain, in this case *Escherichia coli,* by gentamicin and Cifen.

A study carried out in Benin on the chemical composition and *in vitro* efficacy test of essential oils extracted from fresh leaves of *Ocimum basilicum* and *Ocimum gratissimum* on *Salmonella enterica* serotype Oakland and *Salmonella enterica* serotype Legon proved that the essential oils of these two species have a bactericidal effect on the two strains of *Salmonella enterica* because their CMB/CMI ratio is respectively equal to 4 and 2 for each serotype (kpodekon et *al..,* 2013); whereas the essential oil of *O. gratissimum* in the present study showed a bacteriostatic effect (MBC/MIC=16) on the *Salmonella typhimurium* ATCC: 14028 strain and that of *O. basilicum* showed only an inhibitory effect against this strain at a low concentration (0.1793 mg/mL).

In a study conducted in Burkina Faso to determine MICs of *Ocimum basilicum* essential oil against several bacteria, Bassolé *et al* (2010) determined a MIC of 5 mg/ml for *S. enterica* CIP 105150 and *S. enterica* CIP 105150. *typhimurium* ATCC 13311; 8.3 mg/ml for *Escherichia coli* CIP 105182 and 2.5 mg/ml for *Staphylococus aureus* ATCC 9144, while the MIC of the essential oil of *Ocimum basilicum was 5 mg/ml for S. enterica CIP 105150 and S. typhimurium* ATCC 13311; 8.3 mg/ml for *Escherichia coli* CIP 105182 and 2.5 mg/ml for *Staphylococus aureus* ATCC 9144. *basilicum* in this work on S. *typhimurium* ATCC: 14028 is 0.179mg/mL and for *Escherichia coli ATCC:* 25922: 0.143 mg/mL and for *Staphylococus aureus ATCC* 25923 is 1.147 mg/mL.

The essential oil of *O. basilicum* in our study has a greater inhibitory effect than that of Burkina Faso for the strains common to these two studies and this difference would be due to the serotype of each bacterium and the majority compounds identified in each oil.

Bassolé *et al.* 2001 further evaluated the antibacterial activity of essential oils of *Cymbopogon proximus* and *Ocimum canum* Sims on some Gram- (*Escherichia coli, Proteus mirabilis, Salmonella Enterica, Shigella dysenteria*) and Gram+ (*Bacillus cereus, Enterococcus faecalis, Listeria innocua, Staphylococcus aureus, Staphylococcus camorum*) strains. During their study, it was found that Gram+ bacteria are more sensitive than Gram- bacteria to essential oils and that the essential oil of the leaves of *Ocimum canum* inhibits all germs with inhibition zones varying from 6 to 22 mm and that *Bacillus cereus* and *Staphylococcus camorum* are the most inhibited and *Shigeffa dysenteria* is the least sensitive.

They have also noticed that most Gram- bacteria are less sensitive to this essential oil as is the case with our *Ocimum canum* Sims.

Our results would also be similar to those of Hassan and his collaborators who demonstrated the antibacterial activity of two essential oils of *Ocimum canum* Sims from two sources on the island of Grande Comore and who inhibited *Escherichia coli* and *Staphylococcus aureus* (two ATCC strains) only at their highest concentration (Hassan et *al.,* 2011).

This rapprochement would be due to the fact that the essential oils of *Ocimum canum from* two sources: the island of Grande Comore and that of the DRC contain the same majority component, namely 1-8-cineole.

Ramanoelina et al (1987) showed that the essential oil of *Ocimum gratissimum* had antibacterial activity on several reference strains including: *Staphylococus aureus* ATCC6538 on which it presented an MIC=0.416 mg/mL), *Klebsiella pneumoniae* ATCC1003 on which this same oil was active with MIC=0.250 mg/mL and *Escherichia coli* IPP54127 which was inhibited at MIC=0.416mg/mL of this oil. The MICs of the essential oil of *O. gratissimum* in our work on the same strains but other serotypes are high: 2.01-4.02mg/mL. This difference in concentrations would be due, as in other cases, to the difference in serotype and perhaps to the difference in the chemical composition of the two oils.

This same study (Ramanoelina *et al.*, 1987) was also carried out on several other wild strains including *Escherichia coli , Proteus mirabilis* and *Proteus rettgeri and it was concluded* that the essential oil of *Ocimum gratissimum* did not act on the two strains of *Proteus* whereas the essential oil of the species of *Ocimum gratissimum from* DRC inhibited Proteus *mirabilis,* an ATCC strain at MIC=0.125mg/mL.

Although Gram+ germs constitute a large part of the pathogenic germs, and most essential oils have been shown to inhibit these germs, the main criterion for the sensitivity of strains seems to be Gram-, being in general, more resistant than Gram+ bacteria (Ramanoellina *et al.,* 1987).

The spectrum of action of our two essential oils (*Ocimum basilicum* and *Ocimum gratissimum*) seems broad because they inhibit the Gram+ and Gram- germs on which they have been tested.

It is noted that among the microbial strains developed by sickle cell disease patients, most are Gram-positive, which are often resistant to many essential oils. The three species of *Ocimum* studied would constitute essential oil resources for the aromatherapy of these patients alongside their anti sickle cell disease activity.

3.3.3. Anti-radical activity of extracts and essential oils

Work has shown that there is a correlation between the proportion of irreversible sickle cell and membrane damage of red blood cells and that when the rate of irreversible sickle cell is less than 5% in sickle cell patients, the membrane characteristics are close to those of GR of normal subjects. On the other hand, when the number of sickle-cells is higher than 5%, the level of lipoperoxidation products is higher; thus He concluded that membrane damage caused by oxidative stress is a secondary factor that perpetuates the formation of irreversible sickle-cells (Nacoulma *et al.*, 2006b).

It was therefore important to see the *in vitro* antiradical activity on the DPPH radical of different extracts and essential oils of these three plants using ascorbic acid as a positive control

Thus, the IC50 and 1/IC50 parameters of the extracts and essential oils of the leaves of three plants, compared to those of ascorbic acid are grouped in table 3.39 below, knowing that a low value of IC50 indicates a high antioxidant activity.

Table 3.39: Antioxidant Activity of Ascorbic Acid, Extracts and Essential Oils

Plant	Extract	IC50 mg/mL	1/IC50	Observation
O. basilicum	Methanol	0,020	50,00	Very high, >AA
	Ethyl acetate	0,090	11,11	high
	Dichloromethane	-	-	-
	n-hexane	-	-	-
	Essential oil	1,18	0,84746	Low
O. canum	Methanol	0,0234	42,7350	Very high
	Ethyl acetate	0,0767	13,0378	High
	Dichloromethane	0,1938	5,1599	average
	n-hexane	-	-	-
	Essential oil	1,4217	0,7034	Low
O. gratissimum	Methanol	0,0576	17,3611	High
	Ethyl acetate	0,0508	19,6850	High
	Dichloromethane	0,0787	12,7065	High
	n-hexane	0,1345	7,4349	average
	Essential oil	0,0389	25,7069	Very high
	Thymol	0,127	7,874	Average
Positive Control	Ascorbic acid	0,0220	45,4545	Very high

Knowing that the antiradical activity is: Low: $0 < 1/IC50 < 5$; Medium: $5 < 1/IC50 < 10$; High : $10 < 1/IC50 < 20$; Very high : $1/IC50 \geq 20$

It can be seen from this table that the methanolic extract of *O. basilicum* is the most active (1/IC50 =50) of all the extracts and its activity even exceeds that of the positive control (AA). Its ethyl acetate extract shows a high anti-radical activity while its two apolar extracts (ED and EH) showed no activity. However, the essential oil of *O. basilicum* showed low antioxidant activity (Tshilanda *et al.*, 2016a).

It is also noted that the methanolic extract of *O. canum* has a very high antiradical activity with 1/IC50=42.7 close to that of ascorbic acid taken as a control (1/IC50: 45.45); its ethyl acetate extract also has a high activity. However, its DCM extract (apolar) shows a medium antiradical activity while its essential oil has a low antiradicalare activity.

From Table 3.39. above, it is also observed that among the four extracts studied of *O. gratissimum*, the ethyl acetate extract has a high antiradical activity (1/IC50:19,685) followed by the methanolic extract (1/IC50:17,3611). On the other hand, its two apolar extracts (ED and EH) exhibited high and medium *activity* respectively and its essential oil has a very high activity (1/IC50:25,7069) which is higher than that of all the extracts.
As in the cases of *Ocimum canum* and *O. basilicum*, polar extracts of *Ocimum gratissimum* have a greater antiradical activity than apolar extracts.
Thus, *O. gratissimum* can be considered as the most active plant by the fact that it is the only species whose extracts (polar and non-polar) show antiradical activity and whose essential oil has a higher antioxidant activity than all the extracts of the same plant, whereas the essential oils of *O. canum and O. basilicum* show a low antiradical activity.

The difference between the IC50 values of methanolic and acetic extracts of two species: *O. canum* and *O. basilicum* is quite large (0.0234mg/mL-0.0767m/mL) and (0.020 mg/mL- 0.09 mg/mL) respectively while the difference between the IC50 values of methanolic, acetatic and dichloromethane extracts of *Ocimum gratissimum* is not large (0.0576mg/mL-0.0508mg/mL-0.0787mg/mL).

.

Only the methanolic extract of *O. basilicum* has an IC50 value (0.020 mg/mL) slightly lower than that of the positive control (0.022 mg/mL) and close to that of the IC50 of the methanolic extract of *O. canum (0.*0234 mg/mL) indicating that the methanolic extracts of these two plants are more active than that of *O. canum. gratissimum* (0.0576 mg/mL); on the other hand, the acetate extract of O. *gratissimum* has a low IC50 (0.0508 mg/mL) compared to the acetate extracts of *O. canum (0.0767* mg/mL) and *O. basilicum (0.09* mg/mL) indicating that this extract is more active than its counterparts from two other species.

Similarly, the DCM extract of *O. gratissimum* (0.0787 mg/mL) showed greater antiradical activity than its counterpart of *O. canum (*0.1938 mg/mL) and the acetate extract of O. *basilicum.*

The hexanic extract of *O. gratissimum* also exhibited superior antiradical activity compared to the DCM extract of *O. canum.*

Although [the] methanolic extract of *O. canum* and *O. basilicum* have a high antiradical activity on DPPH. (which approaches that of ascorbic acid taken as a control) compared to that of *Ocimum gratissimum*, all extracts of the latter are active and show low IC50 when compared to their correspondents of two others.

For both plants (*O. basilicum* and *O. canum*), the active extracts exhibited 1/IC50 greater than those of their corresponding essential oils thus implying a greater activity than that of their corresponding essential oils, while for *O. gratissimum* the essential oil exhibited greater anti-radical activity than all four extracts.

Studies conducted on the ethanolic extract and essential oil of the roots of *Vetiveria zizanioides* acclimatized in Togo, have shown that the ethanolic extract is more active than the essential oil of the same plant (Dotsé et *al.,* 2011).

Ocimum gratissimum is the only species whose *n-hexane* extract (EH) has shown antioxidant activity (1/IC50 of EH=7.43).

Antioxidant activity was observed with the essential oils of three species, however only the essential oil of *Ocimum gratissimum* showed better activity.

Indeed Oboh *et al.* in 2008 showed that polar and apolar extracts of *O. gratissimum* are endowed with antioxidant properties but a study comparing the activity of these extracts to that of the essential oil of the same species has not yet been carried out, so it appears that it is for the first time that it is carried out (Tshilanda *et al.,* 2016a).

Another study also demonstrated the antioxidant activity of the methanolic extract of *O. gratissimum* (Akinmoladun et *al.,* 2007).

The difference in the chemical composition of the essential oils and extracts of these three plants would justify the difference in activity observed.

The antiradical activity of the essential oil of *O. canum* would be partly due to that of its major component which is 1,8-cineole (Ciftci et *al.,* 2011).

On the other hand, the essential oil of *O. basilicum* from Algeria with a linalool chemotype showed a low activity (IC50: 83mg/mL) (Khelifa et *al.*, 2014).

Other studies have also shown that the essential oil of *O. basilicum* has low antiradical activity compared to butylated hydroxy toluene (positive control) (Politeo *et al.* 2007; Hussain *et al.* 2008).

There is a correlation between the antioxidant capacity of essential oils and their content of phenolic compounds (Agbodan et *al.*, 2014); this would make it possible to justify the great activity that the essential oil of *O. gratissimum* possesses and this activity would be due, among other things, to thymol which is the major constituent of the essential oil of *O. gratissimum in* our country. This was also proven during a study evaluating the antioxidant activity of the essential oil of *O. gratissimum* with thymol chemotype (50%) which presented a capacity of trapping DPPH radicals attributed to this phenolic compound (Tchobo et *al.*, 2014).

The activity of the methanolic extract of these three plants would be due to the presence of rosmarinic acid (Tshilanda et *al.*,2016b, Rerhow et *al.*,2012, Costa, 2012) and for *O. canum* to the synergy between rosmarinic acid and the other unidentified compound that is in appreciable quantity in this extract when the HPLC chromatogram is observed (Figure 3.2) while this compound is very weak in *O. gratissimum* and *O. basilicum* (Tshilanda et *al.*, 2016b).

The activity of the essential oil of *O. gratissimum* would be due to its composition rich in oxygenated aromatic hydrocarbons including thymol, its majority compound known for its antioxidant activity, carvacrol and (Z)-tert-butyl-4-hydroxyanisole which are powerful antioxidants used in the food, cosmetic, pharmaceutical industries and for the protection of oils and fats (Walter *et al.*, 1983; Rao *et al.*, 1984; Yanishliea et *al.*, 1999).

Certain compounds isolated and identified in some extracts of *Ocimum gratissimum* in this work would also justify their antiradical activity:

- Methanolic extract: Rosmarinic acid, 1,8-dimethoxy-1,8-methyl-3-anthraquinone,...
- Ethyl Acetate Extract: ursolic acid, oleanolic acid
- Dichloromethane extract: oleanolic acid and squalene

These three plants showed a promising antiradical effect *in vitro.*

Thus, the capacity of the extracts and essential oils of these three *Ocimums* to display such a pharmacological property may represent a rational explanation to promote the use of these plants to relieve several diseases causing oxidative stress in general and sickle cell disease in particular and as natural antioxidants in the food, cosmetic and pharmaceutical industries (Tshilanda et al., 2015a, Tshilanda *et al.*, 2016a).

The fractionation of extracts, particularly those of *O. canum* and *O. basilicum in* order to determine the other active molecules is desirable especially to determine the chemical structure of the majority component noted on the HPLC chromatogram for *O. canum* (Tshilanda et *al.*, 2016b), the chemical composition of its essential oil, the chemical structure of DO8 isolated from O. *basilicum*.

Given the good antioxidant activity of the essential oil of *Ocimum gratissimum* against DPPH free radicals and that the majority component of this oil is thymol (39%), a comparative study of their antiradical activities was carried out.

Thus, the results quantitatively confirm that the essential oil of *O gratissimum* has an antioxidant activity and therefore a higher capacity to trap radicals than thymol. The lower IC50 value corresponding to the higher activity leads to the assertion that ascorbic acid is more active, followed by the essential oil and finally thymol, showing that apart from thymol, there are other compounds that confer this activity to the essential oil, thus acting in synergy.

The antifalcemic, antibacterial and antioxidant properties of the extracts, isolated products and essential oils of these three plants are for the most part interesting, and demonstrate the importance of using these three species growing in our country as candidate antifalcemic agents, to treat infectious diseases and those resulting from oxidative stress.

Since sickle cell disease includes apart from the falciformation of erythrocytes, infections and free radicals, these three species can help to relieve sickle cell disease patients.

CONCLUSION AND SUGGESTIONS

This work is part of our research program on Congolese medicinal plants, especially the genus *Ocimum*.

In this work, we set ourselves the objective of determining the chemical composition and evaluating the biological activities of extracts, essential oils and isolated products of *Ocimum basilicum, O. canum* and *O. gratissimum*.

From this study, it was found that these three species growing in the DRC have almost the same chemical profile in most secondary metabolites. The presence of free polyphenols, alkaloids, saponins, triterpenoids, steroids and quinones was noted, but sensitivity to certain reagents is greater for *Ocimum gratissimum* than for the other two species. In addition, *Ocimum canum did* not react to the moss test, which indicates the presence of saponins in a plant; and that as regards phenolic compounds, flavonoids of the kæmpferol type and phenolic acids (including rosmarinic acid) were identified in all three plants studied but in high content in *O. basilicum* followed by O. *gratissimum and* then finally O. *canum*.

It thus appears that this is the first time that a comparative study of the chemical profile of these three plants growing in our country has been carried out (Tshilanda *et al.,* 2016b).

It also appears that some compounds have been isolated from *O. gratissimum* and their structures determined. These are ursolic acid, oleanolic acid and squalene. In addition, 35 compounds contained in the same plant were identified, including cyclohexatetracontane, 1,8-dimethoxy-3-methyl-3-anthraquinone and eicosanol.

35 compounds have been identified in the essential oil of *O. basilicum of* which the main ones are methylchavicol (35.72%), linalol (21.25%) and in the essential oil of *O. gratissimum* 40 compounds of which thymol (39.86%), γ-terpinene (22.34%) and *p-cymene* (17.60%) are the majority. A majority compound, 1,8-cineole, in the essential oil of *O. canum* has also been identified.

It is noticed after this study that the essential oil of *O. gratissimum from* our country contains a rare isomer of limonene, sylvestrene. This is the first time that this constituent is found in this oil.

To our knowledge, for plants growing in the DRC, it is also the first time that such a large number of chemical compounds have been determined from *O. gratissimum*, that an intensive study of the chemical composition of the essential oils of O. *basilicum and O. gratissinum has* been carried out and that a study of the chemical composition of the essential oil of *O. canum has been* initiated.

At the end of this study, it is therefore obvious to distinguish *Ocimum basilicum* (which is the species growing in Kinshasa and Central Congo) from *O. canum* (the species harvested in Mbuji-Mayi) by their chemotypes, namely: methylchavicol-linalol for *Ocimum basilicum* and 1,8- cineole for O. *canum. canum* and remove the ambiguity of the names: O. *basilicum, O. americanum* (or *canum*) attributed to the only species growing in Kinshasa and Central Congo as indicated by Lathan & Konda Ku Mbuta (2007) and thus contribute to modernize the systematics of the species of the genus Ocimum.

A few non-aromatic esters with sickle cell activity have been synthesized. These include esters of palmitic acid, stearic acid and oleic acid.

Pharmacological screening showed that all plants tested had anti-sickle cell properties, which are due to organic acids (rosmarinic acid, ursolic acid, oleanolic acid) and anthocyanins (Tshilanda et al., 2016b, Tshilanda *et al.*, 2015b; Tshilanda et *al.*, 2015c).

The essential oils extracted from these plants have also shown an anti sickle cell disease activity, thymol being the active principle for the essential oil of *O. gratissimum*.

In addition, the resulting non-aromatic esters have shown *in vitro* antifalcemic activity due to their hydrophobic properties that promote cell membrane permeation; this activity was not observed for oleic acid esters (Tshilanda et *al.*, 2016c).

To our knowledge, it is for the first time that the antifalcemic activity of these plants, its essential oils, rosmarinic acid, ursolic acid, thymol and non-aromatic esters have been studied and observed.

The essential oil of *O. gratissimum* has shown a bactericidal effect on two strains (*Proteus mirabilis* and *Pseudomonas aeruginosa*) and a bacteriostatic effect on *Escherichia coli, Staphylococus aureus, Klesbiella pneumonia and Salmonella typhimurium.*

It also appears from this study that the three plants have an anti-free radical activity, of which that of *O. gratissimum* is superior by the wide field of action of its extracts and essential oil.

It is noted that this is the first time that a comparative study of the antioxidant activity of the extracts and essential oils of these three plants has been carried out.

Thus, the anti-sickle cell, antibacterial and anti-radical properties of these aromatic plants make them interesting in the formulation of drugs for a better management of sickle cell disease. This indicates at the end of this study that *O. gratissimum* and *O. canum can* be incorporated into the Congolese pharmacopoeia against sickle cell disease.

From the above we believe we have completely achieved the goal of this work.

We suggest:

- That studies be continued on the methanolic extract of *O. canum in* order t o determine   the structure of the majority compound revealed by HPLC ;

- That, by experimental or theoretical methods, the modes of action of isolated molecules be determined with a view to incorporating them into the management of sickle cell disease.

- That consideration be given to obtaining by synthesis the active molecules isolated from these plants and their homologues to test their activity in order to broaden the range of active products.

- That the procedure be considered for the incorporation of these plants as food supplements not only for sickle cell disease patients but also for everyone in order to prevent certain infections and other diseases due to oxidative stress.

- That aromatherapy is recommended in medicine;

- That popularization be done so that everyone knows his or her type of hemoglobin in order to be informed in the choice of partner before marriage and thus avoid sickle cell disease, this genetic and fatal disease.

- That the Congolese state should take a real interest in research, which is the driving force behind a country's development and the visibility factor of university institutions, by providing them with appropriate and modern equipment and financing them.

BIBLIOGRAPHY

1. Adams RP. Identification of Essential Oil Components by Gas Chromatography Mass Spectroscopy. Allured Publishing Corporation. 2007; 4th Edition. Carol stream. Illinois, USA.

2. AFA. French Adoption Agency. Sickle cell disease: mode of transmission sheet. Available: http://agence-adoption.fr/home/IMG/pdf/LA_DREPANOCYTOSE.pdf. Accessed: 22-09-2015.

3. AFSS. Agence française de sécurité sanitaire des produits de santé: the introduction of recommendations on quality criteria for essential oils (May 2008). http://www.ansm.sante.fr/var/ansm_site/storage/original/application/657257784ff10b1665 4e1ac94b60e3fb.pdf accessed May 2015

4. Agbodan KA, Dotsè K, Koumaglo KH. Antioxidant activities of the essential oils of three aromatic plants acclimatized in Togo. Int. J. Biol. Chem. Sci. 2014; 8(3): 1103-1110. ISSN 1997-342X, http://dx.doi.org/10.4314/ijbcs.v8i3.23

5. Agbor AGL, Kotué TC, Mouotsouo JP, Nanfack P, Nkam M, Ngogang YJ. "Hemodya": a phytomedicine for sickle cell disease management in Cameroon. A. World Journal of Pharmaceutical Research SJIF. 4 (1): 102-112

6. Akinmoladun AC, Ibukun EO, Emmanuel A E, Obuotor EM, Farombi EO. Phytochemical constituent and antioxidant activity of extract from the leaves of *Ocimum gratissimum*. Sci. Res. Essays. Mai 2007; 2 (5): 163-166.

7. Ali ME, Shamsuzzaman LAM. Investigations on Ocimum gratissimum L. III, Constituents of the essential oil. Sci. Res. (Dacca). 1968; 5: 91-94.

8. Anonyme 6. Final report on the safety assessment of squalane and squalene. Journal of the American College of Toxicology. 1982; 1(2): 37–56 (DOI 10.3109/10915818209013146

9. Anonymous 1. Basil (Plant) Ocimum basilicum. Available: http://fr.wikipedia.org/wiki/Basilic_(plant) accessed March 10, 2013.

10. Anonymous 4. Essential oils. Available: http://fr.ekopedia.org/Huile_essentielle. Accessed January 01, 2012.

11. Anonymous 2. Ocimum gratissimum. Available : http: //www.africa-plants.com/1_ocimum_gratissimum_medi.htm. Accessed March 12, 2013

12. Anonymous 3. Ocimum canum. Agro Products. Available: http:// www.agriculturalproductsindia.com/arromatic . Accessed March 12, 2013.

13. Anonymous 5. Terpenes. Available: http://fr.wikipedia.org/w/index.php?title=Terpeneldid=113636814 . Accessed on 08 May 2015

14. Anthony C. Squalene and Squalane Emulsions as Adjuvants Methods. Allison. 1999; 19(1): 87-93.

15. Avlessi F, Dangou J, Wotto DV, Alitonou G.A, Sohounhloue DK, Menut C. Antioxidant properties of the essential oil of the leaves of *Clausena anisata* (Wild) Hook. C. R. Chemistry. Oct-Nov 2004; 7 (10-11): 1057-1061.

16. Babady B. Course of natural substances. Second Bachelor's degree in Chemistry. University of Kinshasa. Kinshasa. 1986.

17. Babalola IT, Shode FO. Ubiquitous Ursolic Acid: A Potential Pentacyclic Triterpene Natural Product. Journal of Pharmacognosy and Phytochemistry. 2013; 8192(2): 214. www.phytojournal.com

18. Babushok VI, Linstrom PJ, Zenkevich IG. Retention Indices for Frequently Reported Compounds of Plant Essential Oils. J. Phys. Chem. Ref. Data **40**, 043101. 2011; http://dx.doi.org/10.1063/1.3653552

19. Barbosa LCA, Demuner AJ, Clemente AD, Paula VF, Ismail FMD. Seasonal variation in the composition of volatile oils from Schinus terebintifolius RADDI. Quim Nova. 2007; 30:1959–1965.

20. Baritaux O, Amiot M-J,Nicolas J. Enzymatic browning of basil (ocimum basilicum L.) studies on phenolic compounds and polyphenol oxidase. Science des aliments. 1991; 11 (1) 49-62.

21. Bassolé IHN, Ouattara AS, Nebier R, Ouattara CAT, Kabore Z, Traore SA. Chemical composition and antibacterial activities of the essential oils of the leaves and flowers of Cymbopogon proximus (Staph) and Ocimum canum (Sims). Pharm. Méd.Trad. AF. 2001; 11:37-51.

22. Bassolé IHN, Lamien-Meda A, Bayala B, Tirogo S. Composition and Antimicrobial Activities of Lippia multiflora Moldenke, Mentha x piperita L. and Ocimum basilicum L. Essential Oils and Their Major Monoterpene Alcohols Alone and in Combination. Molecule. 2010; 15(11): 7825-7839; doi:10.3390/molecules15117825

23. Beauchamp P. Basic Mass Spectroscopy. 2013
@;y: files\classes\spectroscopy home\spectroscopy workbook, latest MS full chapter.doc, California State Polytechnic University, Pomona, USA.

24. Bendiabdellah A, Dib MA, Djabou N, Allali H, Tabti B, Muselli A, Costa J. Biological activities and volatile constituents of *Daucus muricatus* L. from Algeria. Chemistry Central Journal. 2012; 6:48.

25. Berhow MA, ObengAffum A, Gyan B A. Rosmarinic Acid Content in Antidiabetic Aqueous Extract of *Ocimum canum* Sims Grown in Ghana. J Med Food. 2012; 15 (7): 611– 20.

26. BMRD. Biological Magnetic Resonance Data Bank. Disponible: http://www.bmrb.wisc.edu/metabolomics/metabolomics_standards.shtml Mars 2014.

27. Borchman D, Yappert MC, Milliner SE, Smith RJ, Bhola R. Confirmation of the presence of squalene in human Eylid by heteronuclear single quantum correlation spectroscopy. Lipids.2013; 48: 1269-1277. Doi 10.1007/s11745-013-3844.9

28. Bourgeais A and Florentin E. Data sheet on column chromatography technique. File Experimental chemistry, analysis and characterization. Published on Nov. 25, 2002 on the site: http://culturesciences.chimie.ens.fr/taxonomy/term/2741424344

29. Breitmaier E. Terpenes: Flavors, Fragrances. Pharmaca, Pheromones. 2006; 1st Edition Wiley-VCH, Weinheim.

30. Bruneton,J. Elements of Phytotherapy and Pharmagognosia. Paris. 1987; 2: 250-256.

31. Bruneton J. Pharmacologie, phytochimie, plantes médicinales, 3rdEdition, Tec et Doc Paris. 1999 ; Available: http://fr; wikipedia.org/wiki/polyph%C3%A9nol#cite_note-10.Consulted 16/02/20011.

32. Bruneton J. Pharmacognosy - Phytochemistry, medicinal plants. 4th ed, revised and expanded, Tec & Doc - International Medical Editions. Paris.2009; 1288 (ISBN 978-2-7430-1188-8

33. Buck MH. The Nagoya Protocol on Access to Genetic Resources and the Fair and Equitable Sharing of Benefits Arising from Their Utilization to the Convention on Biological Diversity. Review of European Community & International Environment Law. 2011; 20: 47-61. http://dx.doi.org/10.1111/j.1467-9388.2011.00703.x

34. Burt S. Essential Oils: their antimicrobial properties and potential applications in foods: a review. International Journal of Food Microbiology. 2004; 94: 223-253.

35. Catonne Y, Mukasa M, Rouvillain JL, Ribeyre D. Osteo-articular manifestations of sickle cell disease. International Day of Sports Medicine 11-18, Lyon Hip Days. Orthopaedic Journal. Lyon. 16 Oct.2012.

36. Chagonda LS, Makanda CD, Chalchat JC. The essential oils of Ocimum canum Sims (basilic camphor) and Ocimum urticifolia Roth from Zimbabwe. Flav. Fragr. J. 2000; 15: 23–26.

37. Chalchat JC, Garry RP, Sidibe L.   Aromatic plants of Mali. Ocimum basilicum and Ocimum canum Sims. Riv. Ital. EPPOS (spec. Num.). 1996; 7: 618–626.

38. Chalchat JC, Garry R-P, Sidibe L, Harama M.   Aromatic plants of Mali (II). Chemical composition of essential oil of Ocimum canum Sims. J. Essent. Oil Res. 1999; 11: 473–476.

39. Chalchat JC, Özzcan MM. Comparative essential oil composition of flowers, leaves and stems of basil (Ocimum basilicum L.) used as herb. Food Chemistry. 2008; 110(2): 501-3.

40. Charles J Pouchert; Jacqlynn Behnke; Aldrich Chemical Company. Ed. 1. Milwaukee: Aldrich Chemical Co. 1993

41. Charles DJ, Simon JE. A new geraniol chemotype of O. gratissimum L. J. Essent. Oil Res. 1992; 4: 231-234.

42. ChEBI: Chemical   Entities   of   Biological   Interest http://www.ebi.ac.uk/chebi/advancedSearchFT.do consultee Avril 215.

43. Chen HM, Lee MJ, Kuo CY, Tsai PN, Liu JY, Kao Sh. *Ocimum gratissimum* Aqueous Extract Induces Apoptotic Signalling in Lung Adenocarcinoma Cell A549. Evidence Based complementary Alternative Medicine. 2011; 2011: 1-7. Article ID 739093. http://dx.doi.org/10.1155/2011/739093

44. Cherbuliez E, Baehler Br, Rabinowitz J. (1964), Etude de structures peptidiques à l'aide du phénylisothiocyanate VI. Thin layer chromatography of phenylthiohydantoins derived from amino acids. 1964 ; 47(5) : 1350-1353. Article first published online: 24 OCT 2004, DOI: 10.1002/hlca.19640470530

45. Choudhary R, Kharya MD, Dixit VK, Varma KC. Role of phytohormone on the cultivar and essential oil of Ocimum canum Sims. A potential source of citral. Indian Perf. 1989; 33: 224–227.

46. Ciftci O, Ozdemir I, Tanyildizi S, Yildiz S, Hakan OH. Antioxidative effects of curcumin, β-myrcene and 1,8-cineole against 2,3,7,8-tetrachlorodibenzo-p-dioxin-induced oxidative stress in rats liver.Toxicol Ind Health. 2011; 27: 447-453.

47. Cimanga K, Kambu K, Tona L, Apers S, De Bruyne T, Hermans N, Totté J, Pieters L, Vlietick AJ. Correlation between chemical composition and antibacterial activity of some aromatic medicinal plants growing in the Democratic Republic of Congo. Journal of Ethnopharmacology. 2002; 79: 213-220.

48. Cornu A, Massot R. Compilation of Mass Spectral Data, Supplement 1&2. Heyden London.1975; 1, 2, 3.

49. Costa, RS, Carneiro BT C, Cerqueira-Lima AT, Queiroz NV, Alcântara-Neves NM, Pontes-de-Carvalho LC, Velozo ES, Oliveira EJ and Figueiredo CA. *Ocimum gratissimum* Linn. and Rosmarinic Acid, Attenuate Eosinophilic Airway Inflammation in an Experimental Model of Respiratory Allergy to *Blomia tropicalis*. International Immunopharmacology.2012; 13:126-134. http://dx.doi.org/10.1016/j.intimp.2012.03.012

50. CYBERLIPID CENTER: http://www.cyberlipid.org/simple/simp0004.htm read 15 Nov 2015

51. Dambolena JS, Zunino MP, López AG, Rubinstein HR, Zygadlo AJ, Mwangi JW, Thoithi GN, Kibwage IO, Mwalukumbi JM, Samuel T, Kariuki ST. Essential oils composition of Ocimum basilicum L. and Ocimum gratissimum L. from Kenya and their inhibitory effects on growth and fumonisin production by Fusarium verticillioides. Innovative Food Science and Emerging Technologies. 2010; 11(2): 410–414.

52. Das B, Baruchel S. The Science behind Squalene (The Human Antioxidant). 2000. http://=http/www.amazon.com/science-Behind-Squalene-Human-Antioxydan/dp/1890412953 13 Mai 2015

53. Davoust P. Ecosociosystemes. *http://www.ecosociosystemes.fr/terpenes.html)* 14 May 2015.

54. De Medici D, Pieretti S, Salvator G, Nicoletti M, Rasoanaivo P. Chemical analysis of essential oils of Malagasy medicinal plants by gas chromatography and NMR Spectroscopy. Flav. Fragr. J. 1992; 7: 275-281.

55. Denys JC and James ES. Comparison of Extraction Methods for the Rapid determination of Essential Oil Content and Composition of Basil. J. Amer. Soc Hort. Sci. 1990; 115(3):458-462.

56. De Viel E. Sickle cell trait: the benefits of sport. Lequotididiendumedecin.fr. 26.04.2012

57. Dewick PM. The Biosynthesis of Shikimate Metabolites. Natural Product Reports. 1995; 12: 579-607.

58. Diagne I, Diagne-Gueye NDR, Signate-Sy I H, Camara B, Lopez-Sall PH, Diack-Mbaye A, Sarr M, Ba M, Sow HD, Kuakuvi N. Management of sickle cell disease in children in Africa, experience of the cohort at the Albert Royer Children's Hospital in Dakar. Med Trop. 2003; 63: 513-520.

59. Diatta A, Sall N, Sarr N, Diallo F, Toure M. Assessment of oxidative stress in sickle cell disease. The Eurobiologist. 1999; 33(241): 57-60. ISSN 0999-5749.

60. Dibong SD, Mpondo Mpondo E, Ngoye A. Vulnerability of wild fruit species sold in markets in Douala (Cameroon) (2011), Journal of Animal & Plant Sciences. 2011; 11(3): 1435-1441. http://www.biosciences.elewa.org/JAPS; ISSN 2071 - 7024.

61. Dorn-Beineke, A and Frietsch, T. Sickle Cell Disease-Pathophysiologie, Clinical and Diagnostic Implications. Clin Chem Lab Med. 2002 ; 40(11):1075-1084.

62. Dotsè k, Agbodan AK, Nebie Ch.H, Koumaglo HK. Chemical composition and antioxidant activity of the essential oil of the roots of Vetiveria zizanioides acclimatized in Togo. Journal of the West African Chemical Society. 2011; (31): 77-83. ISSN 0796-6687. http://www.soachim.org.

63. Duprez JN, Bardiau M. Chemical composition and *in vitro* efficacy test of essential oils extracted from fresh leaves of common basil (*Ocimum basilicum*) and tropical basil (*Ocimum gratissimum)* on *Salmonella enterica* serotype Oakland *and Salmonella enterica* serotype Legon. J. Soc. West-Afr. Chim. 2013; 035: 41 - 48.

64. Dupuis G. Cours de chimie organique. in http://www. faidherbe.org/site/cours dupuis/acides.htm#ac gras. The 10/05/2013

65. Ekundayo O, Laakso I, Hiltunen R. Constituents of the volatile oil from leaves of Ocimum canum Sims. Flav. Fragr. J. 1989; 4: 17–18.

66. Elion J, Laurance S, Lapoumeroulie C, Physiopathology of sickle cell disease. Tropical medicine. 1999; 47(1): 59-64. ISSN 0025-682X CODEN METRA2

67. El-Sayed AM. The Pherobase: Database of Insect Pheromones and Semiochemicals. 2003-2014. http://www.pherobase.com/ consulté Déc. 2014

68. Emokpae MA, Uadia P O, Gadzama AA. Correlation of oxidative stress and inflammatory markers with the severity of sickle cell nephropathy. Annals of African Medicine. 2010; 9(3): 141-6. **DOI:** 10.4103/1596-3519.68363

69. EWG. The Environmental Working Group. Available: http//www.enviroblog.org/2008/02/unilever.takes-a-bit-out-of-your-face-cream.html 13 Mai 2015

70. Encyclopedia of Organic Chemistry. www.larousse.fr/encyclopedie/divers/chimie_organique/75264 13 May 2015.

71. Fun CE, Baerheim Svendsen A. Composition of Ocimum basilicum var. canum Sims and O. gratissimum L. grown on Aruba. Flav. Fragr. J. 1990; 5: 173-177.

72. Galacteros, F. Sickle cell disease centre, in http://www.integrascol.fr/fichemaladie.php. le 24/03/201013

73. Ganuyu O. Antioxidative potential of Ocimum gratissimum and Ocimum canum Leaf polyphenols and protectiveEffects on some pro-oxidants Induced lipid peroxidation in Rat Brain: An in vitro study. Am.J.Food Technol. 2008;3 (5): 325-334.

74. Gazengel, U.M.; Orechioni, A.M. Le préparateur en pharmacie, Guide Théorique et Pratique, techniques et documents, 3rd Edition, Paris, 1999.

75. Gorecki M, Acquaye CTA, Wilchek M, Votanot JR, Richt A. Antisickling activity of amino acid benzyl esters. Proc. Natl. Acad. Sci. USA, Biochemistry. 1980; 77(1):181- 185. http://dx.doi.org/10.1073/pnas.77.1.181

76. Goretti VMS, Vieira IGP, Francisca N P M, Irineu. L A, Rogério Nd S, Fábio O S and Selene Morais SM. Variation of Ursolic Acid Content in Eight Ocimum Species from Northeastern Brazil. Molecules. 2008 ; 13(10) :2482-2487. doi:10.3390/molecules13102482 Communication

77. Guillaumin A, Moreau F, Moreau C. The life of plants. Editions Holliers. 1955; Paris VI, 356-36.

78. Gurib-Fakim A. Medicinal plants: Traditions of yesterday and drugs of tomorrow. Molecular Aspects of Medicine. 2006; 27: 1-93.

79. Harnafi H, Hana Serghini HC, Bouanani N el H, Aziz M, Amrani S. Hypolipemic. Activity of polyphenol-rich extracts from *Ocimum basilicum* in Triton WR-1339-induced hyperlipidemic mice. Food chemistry. 2008; 108 (1): 205-212. Doi : 10.1016/j. food chem.2007.10.062

80. Hassani SOS, Ghanmi M, Satrani B, Farah AF. Amarti SM, Achnet, Chaouch A. Chemical composition and bioactivity of essential oils from two sources of *Ocimum canum* S from the island of Grande Comore. Phitotherapy. 2011; 9:18-24.

81. Hegnauer R. Chemotaxonomy of plants. Birkhauser Publishing House Basel. 1966 ; 4 : 314.

82. Hostettmann K, Poterat O, Wolfender,JL. The potential of higher plants as a source of drugs. Chimia. 1998; 52: 10-17.

83. Hostettmann K, Marston A. Twenty years of research into medicinal plants: results and perspectives. Phytochemistery reviews. 2002; 1: 275-285.

84. Horowitz G. Undergraduate Separations Utilizing Flash Chromatography. J. Chem. Educ. 2000; 77: 263.

85. Hussain AI, Anwar F, Tufail Hussain SS, Przybylski R. Chemical composition, antioxidant and antimicrobial activities of basil (*Ocimum basilicum*) essential oils depends on seasonal variations, Food Chemistry. 2008; 108(1): 986-995 www.sciencedirect.com

86. Ibrahim H, Sani FS, Danladi BH, Ahmadu AA. 2007. Phytochemical and antisickling studies of the leaves of Hymenocardia acida Tul (Euphorbiaceae), Pakistan J. of Biological Sci. 2007; 10(5): 788-791.

87. Ismail Z and Azhari H N. Morphological characterization and essential oil composition of Basil (Ocimum basilicum L.) accessions introduced and growing in Malaysia. UMTAS 2011 Empowering Sci, Techn and Innov Towards a Better Tomorrow. 2011; 1-8.

88. Iyamu EW, Turner EA, Asakura T. In vitro effects of NIPRISAN (Mix − 0690), a naturally occurring, potent antisickling agent. British Journal of Haematology. 2002; 118: 337-343.

89. Jamal J, Khalighi A, Kashi A, Bais HP, and Vivanco M. Chemical Characterization of Basil (Ocimum basilicum L.) Found in Local Accessions and Used in Traditional Medicines in Iran. J. Agric. Food Chem., 2002, 50 (21), pp 5878–5883.DOI: 10.1021/jf020487q

90. James ES, Morales MR, Phippen WB, Vieira RF and Hao Z. 1999. Basil: A source of aroma compounds and a popular culinary and ornamental herb. Janick edition, ASHS. 1999; 499–505.

91. Javanmardi J, Stushnoff C, Locke E, Vivanco JM.ᐧ Antioxidant activity and total phenolic content of Iranian *Ocimum* accessions,.Food chemistry. 2003; 83(4):547-550.

92. Jiménez-Arellanes A, Luna-Herrera J, Cornejo-Garrido J, López-García S, Castro-Mussot ME, Meckes-Fischer M, Mata-Espinoza D, Marquina B, Torres J, Pando RH. Ursolic and oleanolic acids as antimicrobial and immunomodulatory compounds for tuberculosis

treatment. BMC Complementary and Alternative Medicine. 2013; 13:258. http://www.wikiphyto.org/wiki/Acide_ol%C3%A9anolique

93. Kallithraka S, Mohdaly AA, Makris DP, KefalasP. Determination of major anthocyanin pigments in Hellenicnative grape varieties (*Vitis vinifera* sp.): association with antiradical activity. Food comp. anal. 2005; 18(5) : 375 - 386.

94. Kapil A, Sharma S. Anti-complement activity of oleanolic acid: an inhibitor of C3-convertase of the classical complement pathway. J Pharm Pharmacol. Nov 1994; 46(11):922-3. PMID 7897600.

95. Katarzyna JL, Gwyn PJ, David RB, Fred EB, Martin VP, Ting SST, Murray H. Characteristics of Essential Oil from Basil (Ocimum basilicum L.) Grown in Australia. J. Agric. Food Chem. 1996; 44(3) : 877-881. DOI: 10.1021/jf9405214.

96. Kelly GS. Squalene and its potential clinical uses. Alternative Medicine Review: a journal of clinical therapeutic. 1999; 4(1): 29-36. (PMID: 9988781.

97. Kelm MA, Nair M G, Strasburg GM, Dewitt DL. Antioxidant and cyclooxygenase inhibitory phenolic compounds from *Ocimum sanctum* Linn. Phytomedicine. 2000; 7(1): 7-13.

98. Kew Databases , World Checklist of Selected Plant Families consulté le 27/01/2016

99. Khelifa LH, Brada M, F. Brahmi F, Achour D, M.L. Fauconnier ML, Lognay G. Chemical Composition and Antioxidant Activity of Essential Oil of Ocimum basilicum Leaves from the Northern Region of Algeria. Top. J. Herbal Medicine. 2012; 1(2): 25-30. Available online at http://www.topclassglobaljournals.org ISSN 2315-8840 ©2012 Topclass Global Journals.

100.   Khelifa LH, Brada M, Achour D, Wathelet JP, Lognay G. Chemical composition and antioxidant activity of the essential oils of *Ocimum basilicum of* Ain Defla. University publications. 2014; www.univ-chlef.dz/drupub/?q=node/180

101.   Kikufi AB. Phylogenetic classification today. Herbarium edition. DR.Congo. Kinshasa. 2012; 1: 26-27.

102.   Klimankova E, Holadová K, Hajšlová J, Caika T, Poustka J, Koudela M. Aroma profiles of five basil (Ocimum basilicum L.) cultivars grown under conventional and organic conditions. Food chemistry. 2008; 107(1):464-472.

103.   Kpadonou Kpoviessia BGH, Kpoviessia SDS. Ladekan EY, Gbaguidib F, Frédérich M, Moudachiroub J, Quetin-Leclercqc J, Accrombessia GC, Michel J, Bero J. In vitro antitrypanosomal and antiplasmodial activities of crude extracts and essential oils of

Ocimum gratissimum Linn from Benin and influence of vegetative stage. Journal of Ethnopharmacology. 2014; 155: 1417–1423.

104.   Kpodekon MT, Boko KC, Mainil JG, Farougou S, Sessou P, Yehouenou B, Gbenou J, Duprez JN, Bardiau M. Chemical composition and in vitro efficacy test of essential oils extracted from fresh leaves of common basil (Ocimum basilicum) and tropical basil (Ocimum gratissimum) on Salmonella enterica serotype Oakland and Salmonella enterica serotype Legon. J. Soc. West-Afr. Chim. 2013; 035 : 41 - 48

105.   Kpodekon MT, Boko KC, Mainil JG, Farougou S, Sessou P, Yehouenou B, Gbenou J, Kumar AK, Singh P, Tripathi NN. Chemistry and bioactivities of essential oils of some Ocimum species: an overview. Asian Pac. J. Trop. Biomed. 2014; 4(9): 682-694.

106.   Kurt R. Thin Film Chromatography. Print book edition. 2nd ed. rev and increase. Paris: Gauthier-Villars. 1971; 25-32.

107.   Labie D, Elion J. Genetics and pathophysiology of sickle cell disease, in: Sickle cell disease (Girot R, Bégué P, Galacteros,F. eds.). John Libbey Eurotext. 2003; 1: 1-10.

108.   Labie D, Elion J. Molecular and pathophysiological bases of hemoglobin diseases. EMC-Hematology. 2005 ; 2:220-239.

109.   . Labra M, Miele M, Ledda B, Grassi F, Mazzei M, Sala F. Morphological characterization, essential oil composition and DNA genotyping of Ocimum basilicum L. cultivars. Plant Science. 2004; 167(4): 725–731.

110.   Lambert RJW, Skandamis PN, Coote PJ, Nychas GJE. The minimum inhibitory concentration (MIC) of oregano essential oil (OEO) and two of its principle components, ie thymol and carvacrol, against Pseudomonas aeruginosa and Staphylococcus aureus. Journal of Applied Microbiology. 2001;91(3):453–462. DOI: 10.1046/j.1365-2672.2001.01428.x

111.   Lanzetta R, Laonigro G, Parilli M, Breitmaier E. Glycosides from Muscaricomosum. Use of homo- and heteronuclear two-dimensional nuclear magnetic resonance spectroscopy for the structure determination of the novel glycoside muscaroside C. Can J Chem. 1984; 62: 2874-8.

112.   Lathan P, Konda Ku Mbuta A. Useful plants from Bas-Congo, Democratic Republic of Congo. 2nd edition. Kinshasa. DR Congo. 2007; 222-223, ISBN No. 9780955420818.

113.	Laughton M. *et al.*, Inhibition of mammalian 5-lipoxygenase and cyclo-oxygenase by flavonoids and phenolic dietary additives. Relationship to antioxidant activity and to iron ion-reducing ability. Biochem Pharmacol. 1991; 42: 1673-1681 (PMID 1656994).

114.	Lawrence BM (1992). Labiatae oils: mother nature's chemical factor, Essential Oils. Edit., Allured Publ. Corp., Carol Stream, IL. 1992; 1988- 1991.

115.	Lawrence BM. Progress in Essential Oils: Ocimum gratissimum oil, Perfum. Flavor. 1997; 22: 70-74.

116.	Lawrence BM, Powel RH, Peele DM. Variation in the genus Ocimum. Proceeding of the Eighth International Congress on Essential Oils, FEDAROM, Grasse, France.1980; 109–117.

117.	Lederer E. Chromatography in Organic Chemistry and Biology. Vol II, Masson et Cie Editeurs, Paris. 1990; 30-61.

118.	Lee SJ, Umano K, ShibamotoT, Lee KG. Identification of volatile components in basil (Ocimum basilicum L.) and thyme leaves (Thymus vulgaris L.) and their antioxidant properties. Food Chemistry. 2005; 91(1): 131–137.

119.	Lemgoum D, Van Bortel L, Van de Borne P. Vascular aspects of sickle cell disease; Blood-Thrombosis Vessels. John Libbey Eurotext. April 2008; 20(4): 191-6. DOI: 10.168/stv.2008.0269

120.	Lazy, J. Techniques de laboratoire ,Tome I, Fascicule I, Edition III , Masson et Scie, Paris ,1963, 83.

121.	Mahomoodally FM. Traditional Medicines in Africa: An Appraisal of Ten Potent African Medicinal Plants. Evidence-Based Complementary and Alternative Medicine. 2013; 2013: 1-14 Article ID: 617459. http://dx.doi.org/10.1155/2013/617459

122.	Mahuzier G et Hamon M. Abregé de chimie Analytique, Tome2, Méthodes de sépration, Masson, Paris. 1998; 183 : 204,210.

123.	Malumba, M. Phytochemical composition and nutritional evaluation of oil extracted from the pulp of the fruit of Raphia sese. PhD Thesis 2007, 52-53.

124.	Manach C, Scalbert A, Morand C, Rémésy C and Jiménez L. Polyphenols: food sources and bioavailability. Am J Clin Nutr May. 2004; 79(5): 727-747.

125.	Martins AP, Salgueiro L, Vila R, Tomi F, Canigueral S, Casanova J, Proenca-da-Cunha A, Adzet T. Composition of the essential oil of Ocimum canum, O. gratissimum, and O. minimum. Planta Med. 1999; 65: 187-189.

126.   Marzouk AM. Hepatoprotective triterpenes from hairy root cultures of *Ocimum basilicum* L., Z Naturforsh C. 2009; Mar-Apr 64(3-4): 201-9.

127.   Matasyoh JC, Bendera MM, Ogendo JO, Omollo EO, Deng AL. Volatile leaf oil of Ocimum americanum L. Occuring n Western Kenya. Bull. Chem. Soc. Ethiop. 2006; 20(1): 177-180.

128.   Mehanna AS. (2001) Sickle Cell Anaemia and Antisickling Agents Then and Now. Current   Medicinal   Chemistry.   2001;   8:79-88. http://dx.doi.org/10.2174/0929867013373778

129.   Mercola. Squalene: the swine Flu Vaccine's Dirty Little secret exposed August 04, 2009. Avaible: http://articles.mercola.com/sites/articles/archive/2009/08/04/squalene-the-swine-flu-vaccines-dirty-little-secret-exposed.aspx 13 mai 2015

130.   Monnet D, Edjeme NE, Ndri K, Hauhouot-Attoungbre ML, Ahibo H, Sangare A, Yapo AE. Lipoprotein (a) and proteins of the acute phase of inflammation during homozygous sickle cell disease. Clinical biology analgesics. 2002; 60(1) :101-103. John Libbey Eurotext.

131.   Moody JO, Ojo OO, Omotade OO, Adeyemo AA, Olumese PE, Ogundipe OO. Anti-sickling potential of a Nigerian herbal formula (Ajawaron HF) and the Major plant component (Cissus populnea L. CPK). Phytother. Res. 2003; 17: 1173-1176.

132.   Morag CW, Robert A, Nelson JR, Stuart CF. The effects of a cobra venom factor and Ethyl palmitate on the prolongation of survival of heterologous erythrocytes. Yale Journal of Biology and Medicine. 1970; 43: 173-176.

133.   Mordret F and Helme JP. Lipids: Composition and Glyceride Structures. Ann.Nutr.Alim, Paris. 1975; 29:1-47.

134.   Morris K. Techniques of lipidology :Isolation,analysis and Identification of lipids 2nd revised edition; Departement of Biochemistry,University of Ottawa, Ottawa , Elsevier Amsterdam, New-york, Oxford, 1986.

135.   Moulisha B, Kumar GA, Kanti HP. Anti-leishmanial and Anti-cancer Activities of a Pentacyclic Triterpenoid Isolated from the Leaves of *Terminalia arjuna* Combretaceae. Top J.Pharm Res. April 2010; 9(2) :135-140.

136.   Mpiana PT, Mudogo V, Ngbolua KN, Tshibangu DST, Shetonde OM, Mbala MB. In Vitro Antisickling Activity of Anthocyanins from Ocimum basilicum L. (Lamiaceae). International Journal of Pharmacology. 2007a; 3: 371-4.

137. Mpiana PT, Tshibangu DST, Shetonde OM, Ngbolua KN. In Vitro Anti-Drepanocytary Activity (Anti-Sickle Cell Anaemia) of Some Congolese Plants. Phytomedicine. 2007b; 14: 192-195. http://dx.doi.org/10.1016/j.phymed.2006.05.008

138. Mpiana PT, Mudogo V, Tshibangu DST., Kitwa EK, Kanangila AB, Lumbu JBS, Ngbolua KN, Atibu EK, Kakule MK. Antisickling activity of anthocyanins from Bombax pentadrum,Ficus capensis and Ziziphus mucronata: Photodegradation effect. Journal of Ethnopharmacoloy. Decembre 2008; 120 (3): 413–418.

139. Mpiana PT, MudogoV,Tshibangu DST, Ngbolua KN, Manguala PK, Atibu EK, Kakule MK, Makelele LK and Bokota MT. Antisickling Activity and Thermodegradation of an anthocyanin fraction from Ocimum basilicum L.(Lamiaeceae). Comp Bio. Nat. Pro.2010; 3(2):287-295.

140. Mpiana PT, Lombe BK, Mahano AO, Ngbolua KN, Tshibangu DST, Wimba LK, Tshilanda DD, Mushangalusa FK, Muyisa SK. In Vitro Sickling Inhibitory Effects and Anti-Sickle Erythrocytes Hemolysis of Dicliptera colorata C. B. Clarke, Euphorbia hirta L. and Sorghum bicolor (L.) Moench. Open Journal of Blood Disease. 2013; 3: 43-48. http://dx.doi.org/10.4236/ojbd.2013.31009

141. Mpiana PT, Misakabu FM, Tshibangu DST, Ngbolua KN, Mwanangombo DT. Antisickling Activity and Membrane Stabilization Effect of Anthocyanins Extracts from *Adansonia digitata* L. Barks on Sickle Blood Cells. International Blood Research and Reviews. 2014; 2: 198-212.

142. Mudiyanselage SE, Elsner P, Hamburger M, Thiele JJ. Ultraviolet A Induces Generation of Squalene Monohydroperoxide Isomers in Human Sebum and Skin Surface Lipids *In Vitro* and *In Vivo*. Journal of Invetigative Dermatology. 2003; 120(6): 915-922.doi: 10.1046/j.1523-1747.2003.12233.x.

143. Nacoulma EWC, Sawadogo D, Sakandé J, Mansour A, FH, Sangaré A, Sess AED. Influence of fetal hemoglobin (HbF) level on oxidative stress in homozygous sickle cell disease patients living in Abidjan, Côte d'Ivoire. Bull Soc Pathol Exot. 2006; 99 (4): 241-244.

144. Nacoulma EWC, Sakande J, Kafando, Kpowbié ED, Guissou IP. Haematological and biochemical profile of sickle cell disease patients in the stationary phase at the Yalgado Ouedraogo National Hospital in Ouagadougou. Mali Médical 2006a T XXI. 2006b; 1: 8-11, http://malimedical.org/2006/p8a.pdf

145. Nasciment J C, Barbosa LC, Paula VF, David JM, Fontana R, Silva LAM, França R. Chemical composition and antimicrobial activity of essential oils of Ocimum canum Sims. And Ocimum selloi Benth.   Anais da Academia Brasileira de Ciências. 2011; 83(3):787-799.

146. Ndom JC. Contribution to the phytochemical study of two medicinal plants: Crepis cameroonica and Secio burtonii (Asteracees). Biological activities of isolated sesquiterpenoids. PhD thesis. 2008.

147. Neuwinger HD. African traditional medicine: a dictionary of plant use and applications. Mephaim Scientific Publisher, Stutttgart. 2000; **ISBN**3-88763-086-6. Record Number 20000315077

148. Ngassoum MB, Ousmaila H, Ngamo LT, Maponmetsem PM, Jirovetz L, Buchbauer G. Aroma compounds of essential oils of two varieties of the spice plant Ocimum canum Sims from northern Cameroon. J. Food Comp. Anal. 2004; 197-204.

149. Ngbolua KN. Evaluation of the antipanocyte and antimalarial activity of some plant taxa from DR Congo and Madagascar. 2012 PhD thesis, University of Kinshasa, Kinshasa, 300p.

150. Ngbolua KN, Mudogo V, Mpiana PT, Malekani MJ, Rafatro H, Urverg Ratsimamanga S, Takoy L, Rakotoarimana H, Tshibangu DST. Evaluation of the antipanocyte and antimalarial activity of some plant taxa in the Democratic Republic of Congo and Madagascar. Ethnopharmacologia. July 2013; 50: 19-24.

151. Ngbolua KN, Mpiana PT. The Possible Role of a Congolese polyherbal formula (Drepanoalpha) as source of Epigenetic Modulators in Sickle Cell Disease: A Hypothesis. J. of Advancement in Medical and Life Sciences. 2014a; 2(1): 1-3. DOI: 10.15297/JALS.V2I1.02.

152. Ngbolua KN, Bishola TT, Mpiana PT, Mudogo V, Tshibangu DST, Ngombe KN et al. Ethno-botanical survey, in vitro antisickling and free radical scavenging activities of Garcinia punctata Oliv.(Clusiaceae). Journal of Advanced Botany & Zoology. 2014b; 1(2): 1-8.

153. Ngbolua KTN, Bishola TT, Mpiana PT, Mudogo V, Tshibangu DST, Ngombe KN *et al.* Ethno-Pharmacological Survey, In Vitro Anti-Sickling and Free Radical Scavenging Activities of Carapa Procera DC. Stem Bark (Meliaceae). Nova J. Med. Biol. Sci. 2014c; 2(2): 1-14.

154. Ngbolua KN, Tshibangu DST, Mpiana PT, Mazasa SO, Mavakala BK, Ashande MC, Muanishay LC. (2015) Anti-Sickling and Antibacterial Activities of Some Extracts from Gardenia ternifolia subsp. Jovis-Tonantis (Welw) Verdc (Rubiaceae) and Uapaca heudelotii Baill. (Phyllanthaceae). Journal of Advances in Medical and Pharmaceutical Sciences. 2015; 2: 10-19. http://dx.doi.org/10.9734/JAMPS/2015/13427

155. Ngbolua KN, Tshidibi JD, Tshibangu DST, Memvanga PB, Gbolo ZB, Tshilanda DD , Mpiana PT. Drepanoalpha®: An Overview on the Quality Control Process and Standardization Feature of an Antisickling Herbal Drug from Democratic Republic of the

Congo. J. of Modern Drug Discovery and Drug Delivery Research. 2016; 4 (1): 1-6. ISSN: 2348 –3776 homepage: http://scienceq.org/Journals/JMDDR.php

156.	N'Guessan AHO, Déliko CED, Mamyrbékova-Békro JA, Békro IA. Phenolic compound contents of 10 medicinal plants used in the treatment of hypertension, an emerging pathology in Côte d'Ivoire. Journal of Industrial Engineering. 2011; 6: 55-61.

157.	NIST Chemistry WebBook. Available: http://webbook.nist.gov/chemistry/

158.	NIST08, NIST/EPA/NIH Mass Spectral Library (EI)

159.	Nsimba M, Lami N, Hayakawa Y, Yamamoto C. Toshiyuki K (2013). Decreased thrombin activity by a congolese herbal medicine used in sickle cell anemia. Journal of ethnopharmacology. 2013; 148: 895-900.

160.	Nyarko AK, Asare-Anane H, Ofosuhene M, Addy ME. Extract of Ocimum canum lowers blood glucose and facilitates insulin release by isolated pancreatic beta-islet cells. Phytomedicine. May 2002; 9(4): 346-51.

161.	Oboh, G, Raddatz, H, HenleT. Antioxidant properties of polar and non-polar extracts of some tropical green leafy vegetables. Journal of the Science of Food and Agriculture. 2008; 88: 2486–2492. doi: 10.1002/jsfa.3367

162.	Ogunkoya L. Application of mass spectrometry in structural problems in triterpenes. Phytochemistry. 1981; 20:121-126.

163.	WHO, Sickle-Cell anemia, 2006a, pp. 1-5.

164.	WHO. World Health Organization 2006b = WHO, 2006 pages 1-3: Sickle-cell disease and other haemoglobin disorders. 2006; Available: http://**www.who.int/mediacentre/factsheets/fs308/en/** Accessed on 29/01/2016

165.	WHO. World Health Organization. Traditional medicine. 2016; http://www.who.int/mediacentre/factsheets/2003/fs134/fr/ accessed 19/01/2016

166.	Orgogozo JM, Dartigues JF, Lafont S, Letenneur L, Commenges D, Salamon R, Renaud S, Breteler MB. Wine consumption and dementia in the elderly: A prospective community study in the Bordeaux area. Rev. Neurol. April 1997; 153(3): 185-192.

167.	Oswaldo C, Jerome O, Michael RW, Stuart C.F. Survival of Human Sickle-Cell Erythrocytes in: Heterologous Species: Response to Variations in Oxygen Tension. Proc. Nat. Acad. Sci. USA. 1973; 70(8): 2356-2359.

168. Oussou KR, Kanko C, Guessend N, Yolou S, Koukoua G, Dosso M, N'guessan YT, Figureueredo G, Chalchat JC. Antibacterial activities of the essential oils of three plants from Ivory Coast. Chemistry reports. Oct. 2004; 7(10-11): 1081-1086. Doi: 10.1016/j.crci.2003.12.

169. Owokotomo IA, Ekundayo O, Dina O. *Ocimum Gratissimum*: the Brine Shrimps Lethality of a New Chemotype Grown in the South Western Nigeria. Global Journal of Science Frontier Research. 2012; 12(6): 44-49.

170. Ozcan M, Chalchat JC. Essential oil composition of Ocimum basilicum L. and Ocimum minimum L. in Turkey. Czech J. food sci. 2002; 20(6): 223-228.

171. Pandey H, Pandey P, Singh S, Gupta R, Banerjee S. Production of anti-cancer triterpene (betulinic acid) from callus cultures of different *Ocimum* species and its elicidation. Protoplasma. March 2015; 252(2); 647-655. DOI 10.1007/s00709-014-0711-3 Print ISSN0033-183X

172. Patrinos G P,Giardine B, Riemer C, Miller W, Chui DHK, Anagnou NP, Wajcman H and Hardison RC. Improvements in the HbVar database of human hemoglobin variants and thalassemia mutations for population and sequence variation studies., 2004; http://globin.cse.psu.edu/hbvar/menu.html. Le 05/04/2010.

173. Perera P, Andersson R, Bohlin L, Andersson C, Li D, Owen NL, Dunkel R, Mayne CL, Pugmire RJ, Grant DM, Cox PA. Structure determination of a new saponin from the plant *Alphitonia zizyphoides* by NMR spectroscopy. Magn Res Chem 1993; 31: 472-80.

174. Philip M. The use of the stable free radical diphenylpicryl-hydrazyl (DPPH) for estimating antioxidant activity. Songklanakarin J.Sci. Tec.2004; 26 (2): 211-219.

175. Politeo O, Jukic M, Milos M. Chemical composition and antioxidant capacity of free volatile aglycones from basil (Ocimum basilicum L.) compared with its essential oil. Food Chemistry. 2007; 101(1): 379–385. doi:10.1016/j.foodchem.2006.01.045.

176. Pouchert CJ, Behnke J. Aldrich Library of ^{13}C and ^{1}H-FTNMR Spectra. Aldrich Chemical Company. Ed. 1. Milwaukee : Aldrich Chemical Co. 1993.

177. PROTA4U Record display: http://www.prota4u.org/protav8.asp?p=Ocimum+americanum Nov. 10, 2015

178. Ragasa CY and Lim K.Secondary metabolites from Schffleraodorata. (2005) Philippian Journal of Science. 2005: 134 (1): 63-67.

179. Rahimi-Nasrabadi M, Gholivand MB, Batooli H. Chemical Composition of Essential Oil from Leaves and Flowering Aerial Parts of Haplophyllum Robustum Bge (Rutaceae). Digest Journal of Nanostructures and Biostructures. 2009; 4(4) : 819-822.

180. Ramanoelina ARP, Terrom GP, Bianchini JP, Coulanges P. Contribution to the study of the antibacterial action of some essential oils extracted from Malagasy plants. Arch. Inst. Pasteur Madagascar. 1987; 53(1): 217-226.

181. Rao S, Lalwani ND, Watanabe TK, Reddy JK. Inhibitory Effect of Antioxidants Ethoxyquin and 2(3)-ferf-Butyl-4- hydroxyanisole on Hepatic Tumorigenesis in Rats Fed Ciprofibrate, a Peroxisome Proliferato-1 M. CANCER RESEARCH. Mars 1984; 44:1072-1076. Downloaded from cancerres.aacrjournals.org on April 11, 2016. © 1984 American Association for Cancer Research.

182. Reible D, Demnerova K. Estimation using Broto's fragmentation method At order Code At contribution log(p):3.40 ED: 0.13 by Broto's method. Eur. J. Med. Chem.- Chim.Theor. 1984; 19(71). Innovative Approaches to the On-Site Assessment and ...https://books.google.cg/books?isbn=1402009569 2002.

183. Rhourri-Frih B. Analysis, classification and characterization of resins of plant origin by chromatography and mass spectrometry. PhD thesis at the Institute of Organic and Analytical Chemistry of Orleans. Nov 2009.

184. Roudmitska E. The perfume. University Press. Edition I. Paris. 1980 ; 43-44. Available : http .www.herbessence//.Consulté on 5/07/2011.

185. Russu IM, Lin AK, Yang CP, Ho C. Molecular basis for the anti-sickling activity of aromatic aminoacids and related compounds : a proton nuclear magnetic resonance. Biochemistry. 1986; 25(4): 808-15.

186. Ruzicka L and Wallach O. History of the isoprene rule In: Croteau R. The Discovery of Terpenes. Pullman, Washington 99164-6340, USA: Discoveries in plant biology. Chap 20: 329-343.

187. Sahouo BG, Tonzibo ZF, Boti B, Chopard C, Mahy JP, N'guessan YT. Anti-inflammatory and analgesic activities: chemical constituents of essential oils of Ocimum gratissimum, Eucalytptus citriodora and Cymbopogon giganteus inhibited lipoxygenase L-1 and cyclooxygenase of PGHS. Bulletin of the Chemical Society of Ethiopia. 2003; 17(2): 191-197.

188. Saliu BK, Usman LA., Sani A, Muhammad NO, Akolade JO. Chemical composition and Anti-Bacterial Activity (OralIsolates) of leaf Essential oil of Ocimum gratissimumL. Grown in North Nigeria. International Journal of Current Research.2011; 33(3): 022-028.

189.   Sanda K, Koba K, Nambo P, Gaset A.   Chemical investigation of Ocimum species growing in Togo. Flav. Fragr. J. 1998; 13: 226-232.

190.   Sang SK, Lapsley K, Jeong WS, Lachance PA, Ho CT, Rosen RT. Antioxidative phenolic compounds isolated from almond skins (Prunus amygdalus Batsch). J. Agric. Food Chem. 2002; 50: 607-609. In Phenolics in Food and Nutraceuticals- 78. https://books.google.cg/books?isbn=0203508734

191.   Sarni-Manchado P, Cheynier V. Les polyphénols en agroalimentaire, Lavoisier, Editions Tec & Doc. 2006; book 398 p. (ISBN 2-7430-0805-9

192.   Scherer R., Godoy H T. Antioxidant activity index (AAI) by the2, 2'-diphenyl-1-picrylhydrazyl method. Journal of Food Chemistry. 2009; 112(3): 654-658.

193.   SDBS: Spectral Database for Organic Compounds: http://sdbs.db.aist.go.jp/

194.   Seebacher W, N. Simic, R. Weis, R. Saf and O. Kunert. Me-30 of the triterpene moiety of the oleanane-type. Magn. Res. Chem. 2003; 41: 636-638.

195.   Seshata. Healing properties of terpenes and terpenoids. 21 Febr. 2014; http://sensiseeds.com//fr/blog/proprietes-curatives-des terpenes-terpenoides/   13 May 2015.

196.   Sess ED, Carbonneau MA, Thomas MJ, Dumont MF, Peuchant E, Perromat A, Le Bras M & Clerc M. First observations on the main plasma parameters of oxidative stress in homozygous sickle cell disease. Bull Soc Path Exot. 1992; 85 (23): 174-179.

197.   Silverstein RM, Webster FX, Kiemle DJ. Spectrometric identification of organic compounds. (2nd ed.) deboeck, John Wiley &Sons. 2005; 245-282.

198.   Skoog DA, West DM, Holler F J. Analytical Chemistry, De Boeck, University of Brussels. 1997 ; 557-587,858-865.

199.   Sofowora EA, Isaac-Sodeye WA, Ogunkoya LO. Isolation and charactesation of an antisickling agent from Fagara zanthoxylloide root. Lioyidia. 1975; 38: 169-171.

200.   Stashenko EE, Jaramillo BE, Martinez JR. Analysis of volatile secondary metabolites from Colombian *Xylopia aromatica* (Lamarck) by different extraction and headspace methods and gas chromatography. Journal of Chromatography A. 2004; 1025 (1): 105-113.

201.   Stenhagen E, Abrahamsson S, McLafferty FW. Registry of Mass Spectral Data. Wiley, New York. 1974.

202.	Strani B, Ghanmi M, Farah A, Aafi A, Fougrach H, Bourkhiss B, Bousta D, Talbi M. Chemical composition and antimicrobial activity of Cladanthus mixtus essential oil. Bull. Soc. Pharm. Bordeaux. 2007; 146:85-96.

203.	Stryer L, Berg JM, Tymoczko JL. 1997. Biochemistry (5th ed.). Flammarion: Paris.

204.	Swami N, Prema T, Wu Q, Faith C. Nicosan: Phytomedicinal Treatment for Sickle Cell Disease. ACS symposium series. 2009; 1021: 263-276.

205.	Takano T. Effects of the ratios of K to Ca in the nutrient solution on the growth, nutrient uptake, essential oil content and composition of basil. Acta Hort. 1993; 331: 129-142.

206.	Tchobo PF, Alitonou GA, Soumanou MM, Baréa B. Bayrasy C, Laguerre M, Lecomte J, Villeneuve P, Souhounhloue D. Chemical composition and ability of essential oils from six aromatic plants to counteract lipid oxidation in emulsions. Journal of the American Oil Chemists' Society. 2014; 91 (3): 471-479.

207.	Tchoumbougnang F, Dongmo PMJ, Sameza ML, Nkouaya EGM , Tiako GBF, Amvam PHZ, Menut C. Larvicidal activity on Anopheles gambiae Giles and chemical composition of essential oils extracted from four plants cultivated in Cameroon. Biotechnol. Agron. Soc. approx. 2009; 13(1): 77-84.

208.	Tropicos.org. Missouri Botanical Garden in: http://www.tropicos.org 08 Jun 2016

209.	Tshibangu DST, Shode FO, Koobanally N, Mudogo V, Mpiana PT, Ngbolua KN. 2011. Antisickling triterpenoids from Callistemon viminalis, Meulaleuca bracteata var. Revolution Gold, Syzygium guineense and Syzygium cordatum. The 14th NAPRECA Symposium and AAMPS Ethnoveterinary Medicine Symposium, 8 th–12th August. International Cente For Insect Physiology and Ecology (ICIPE): Kasarani, Nairobi, Kenya.2011; .296-300 (YS 27).

210.	Tshilanda DD, Mpiana PT, Onyamboko DV, Mbala BM, Ngbolua KN, Tshibangu DST, Bokolo M K, Taba KM, Kasonga TK. Antisickling activity of butyl stearate isolated from Ocimum basilicum (Lamiaceae). Asian Pac J Trop Biomed. 2014; 4(1): 930-935.

211.	Tshilanda DD, Onyamboko DV, Tshibangu DST, Ngbolua KN, Tsalu PV, Mpiana PT. In vitro Antioxidant Activity of Essential Oil and Polar and Non-Polar Extracts of Ocimum canum from Mbuji-Mayi DR Congo. Journal of Advancement in Medical and Life Sciences. 2015a; 3 (3):1-5. ISSN: 2348-294X.

212.	Tshilanda DD, Onyamboko DNV, Babady P B, Ngbolua KN, Tshibangu DST, Dibwe E d F, and Mpiana PT. Anti-sickling Activity of Ursolic Acid Isolated from the Leaves of Ocimum gratissimum L. (Lamiaceae). Nat Prod Bioprospect. 2015b; 5(4): 215–221. doi: 10.1007/s13659-015-0070-6

213. Tshilanda DD, Onyamboko DV, Mwanangombo DT, Tsalu PV, Misengabu NK, Tshibangu DST, Ngbolua KN, Mpiana PT, In vitro Antisickling Activity of Anthocyanins from Ocimum canum (Lamiacea). J. of Advancement in Medical and Life Sciences. July 2015c; 3(2): 1-4. ISSN: 2348-294X DOI: 10.15297/JALS.V3I2.01

214. Tshilanda DD, Babady PB, Onyamboko DV, Muamba CTT, Tshibangu DST, Ngbolua KN, Tsalu PV, Mpiana PT. Chemo-type of essential oil of *Ocimum basilicum* L. from DR Congo and relative *in vitro* antioxidant potential to the polarity of crude extracts. Asian Pac J Trop Biomed. Sept. 2016a; 6(11): 930-935.

215. Tshilanda D , Mutwale P, Onyamboko, D, Babady P, Tsalu P, Tshibangu, D, Ngombe N, Frederich M, Ngbolua K. and Mpiana P. Chemical Fingerprint and Anti-Sickling Activity of Rosmarinic Acid and Methanolic Extracts from Three Species of Ocimum from DR Congo. Journal of Biosciences and Medicines. 2016b; 4: 59-68. doi: 10.4236/jbm.2016.41008.

216. Tshilanda DD, Ngbolua KN, Tshibangu DST, Malumba AM, Mpiana PT. Synthesis and *in vitro* Antisickling Activity of Some non-aromatic Esters. American Chemical Science Journal. 2016c; 11(3): 1-5. XX-XX Article no.ACSJ.22734 ISSN: 2249-0205. (Short communication).

217. Tsumbu CN, Deby-Dupont G, Tits M, Angenot L, Franck T, Serteyn D, Mouithys-Mickalad A. Antioxidant and Antiradical Activities of *Manihot esculenta* Crantz (Euphorbiaceae) Leaves and Other Selected Tropical Green Vegetables Investigated on Lipoperoxidation and Phorbol-12-myristate-13-acetate (PMA) Activated Monocytes. Nutrients. 2011; 3: 818-838. http://dx.doi.org/10.3390/nu3090818

218. USDA: United states Department of agriculture (USDA), Natural Resource conservation service, plants Database; http://plants.usda.gov/core/profil? Symbol=OCCA4.

219. Valtcho DZ, Amber C, Charles LC. Yield and Oil Composition of 38 Basil (Ocimum basilicum L.) Accessions Grown in Mississippi. J. Agric. Food Chem. 2008; 56(1): 241–245. **DOI:** 10.1021/jf072447y.

220. Van Duong, Nguyen. Medicinal Plants of Vietnam, Cambodia, and Laos. Acta Hort. 1993; 331: 120-127.

221. Vlase, L, Benedec, D, Hanganu, D. Damian, G. Csillag I, Sevastre B, Mot AC, Silaghi-Dumitrescu R. and Tilea I. Evaluation of Antioxidant and Antimicrobial Activities and Phenolic Profile for Hyssopus officinalis, Ocimum basilicum and Teucrium chamaedrys. Molecules. 2014; 19:5490-5507. http://dx.doi.org/10.3390/molecules19055490

222. Voet D, Voet JG. 1998. Biochemistry (2nd ed.) De Boeck University: Paris.

223. Vostrowsky O, Garbe W, Bestmann HJ, Maia JGS. Essential oil of alfavaca, Ocimum gratissimum from Brazilian Amazon. Zeit Naturforsoh. 1990; 45: 1073-1076.

224. Wagner H and Bladt S. (1996) Plant Drug Analysis: A Thin Layer Chromatography Atlas. 2nd Edition, Springer-Verlag. 1996; Berlin. http://dx.doi.org/10.1007/978-3-642-00574-9

225. Walter J, Kozumbo WJ, Seed JL, Kensler TW. Inhibition by 2(3)-tert-Butyl-4-hydroxyanisole and Other Antioxidants of Epidermal Ornithine Decarboxylase Activity Induced by 12-O-Tetradecanoylphorbol-13-acetate. American Association for Cancer Research. June 1983; 43: 2555-2559. Downloaded from cancerres.aacrjournals.org on April 11, 2016.

226. Wang J, Jiang Z, Xiang L, Li Y, Ou M, Yang X, Shao J, Lu Y, Lin L, Chen J, Dai Y, Jia L. Synergism of ursolic acid derivative US597 with 2-deoxy-D-glucose to preferentially induce tumor cell death by dual-targeting of apoptosis and glycolysis. Scientific reports. Mai 2014; 4: 1-12. 5006 DOI: 10.1038/srep05006
http://www.nature.com/srep/2014/140516/srep05006/full/srep05006.html

227. WCSP Kew. World Checklist of Selected Plant Families: Royal Botanic Gardens, Kew. Disponible: http://fr.wikipedia.org/wiki/ocimum[#] Kew_ liste. 10 Sept 2012.

228. Weaver DK, Dunkel FV, Ntezurubanza L, Jackson LL, Stock DT. The efficacy of linalool, a major component of freshly-milled Ocimum canum Sims (Lamiaceae), for protection against postharvest damage by certain stored product Coleoptera. Journal of Stored Products Reseach. 1991;27(4):213-220.

229. Weil, J.H. General biochemistry 10 edn, Dunod. Paris. 2005; 51-61.

230. WHO. World Heath Organisation. Washington post article: Global Advisory Committee on Vaccine Safety. Available: http://www.who.int/vaccine_safety/topics/adjuvants/squalene/qustions-and_answers/en/ accessed 10 May 2015.

231. Wiley Registry 10th Edition. Mass Spectral Library.

232. Yamada AN, Grespan R, Freitag AF, Damião MJ, Kummer R, Lucena R, Bersani-Amado CA, Cuman RKN. Evaluation of the potential anti-inflammatory of Ocimum Americanum L. essential oil in different experimental models of acute inflammation. International Journal of Applied Research in Natural Products. 2014; 7 (3): 1-6.

233. Yanishlieva NV, Marinova EM, Gordon MH, Raneva VG. Antioxidant activity and mechanism of action of thymol and carvacrol in two lipid systems. Food chemisty. 1999 ; 64(1) : 59-66.

234. Yayi E, Moudachirou M, Chalchat JC. Chemotyping of three Ocimum species from Benin: O. basilicum, O. canum, O. gratissimum. J. Essent. Oil Res. 2001; 13: 13-17.

235. Yuma PM, Mpiana PT, Bokota MT, Wakenge LB. Muanishay CL, Gbolo BZ, Mathina G, Tshibangu DST, Ngboluua KN. (2013), Study of the antifalcemia activity and thermo- and photo-degradation of anthocyanins from Centella asiatica, Thomandersia hensii and Maesopsis eminii. Int. J. Biol. Chem. Sci. 2013; 7(5): 1892-1901.

236. Yusuf M, Choudhry J, Wahab MA, Begum J. Medicinal Plants of Bangladesh. BCSIR: Dhaka, Bangladsh. 1994; http://indianmedicine.eldoc.ub.rug.nl/root/Y/150268/

237. Yusuf M, Begum J, Mondello L, Stagno d'Alcontres I. Studies on the essential oil bearing plants of Bangladesh. Part VI. Composition of the oil of Ocimum gratissimum. L. Flav. Fragr. J. 1998; 13: 163-166.

238. Zhang JW, Li SK and Wu WJ. The Main Chemical Composition and in vitro Antifungal Activity of the Essential Oils of Ocimum basilicum Linn. var. pilosum (Willd.) Benth. Molecules. 2009; 14(1) : 273-278; doi:10.3390/molecules14010273

239. Zollo PHA, Biyiti L, Tchoumbougnang F, Menut C, Lamaty G, Bouchet P. Aromatic plants of tropical Central Africa. Part XXXII. Chemical composition and antifungal activity of thirteen essential oils from aromatic plants of Cameroon. Flav. Fragr. J. 1998; 13: 107–114.

240. Zollo PHA, Biyiti L, Tchoumbougnang F, Menut C, Lamaty G, Bouchet P. Aromatic plants of tropical Central Africa. Part XXXII. Chemical composition and antifungal activity of thirteen essential oils from aromatic plants of Cameroon. Flav. Fragr. J. 1998; 13: 107–114.

210

ANNEXES

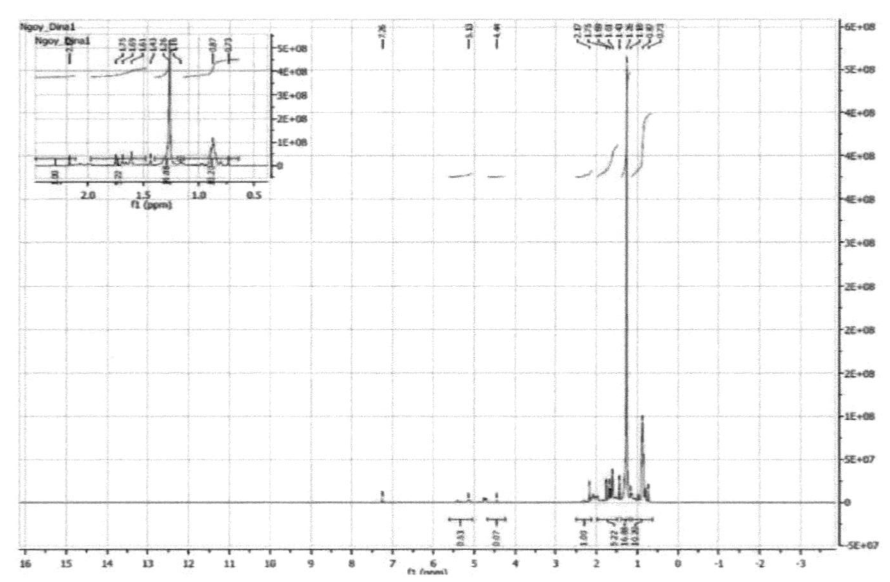

Figure 4.1: 1H-NMR spectrum of Dinal

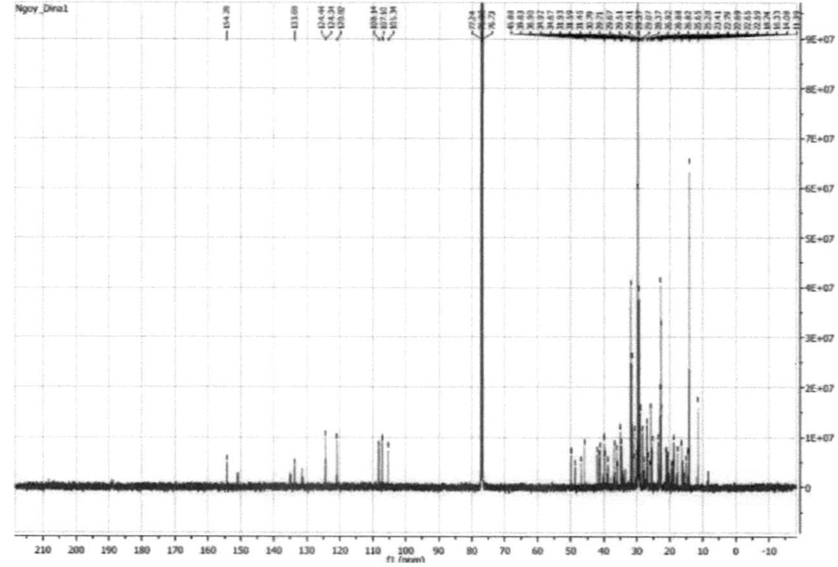

Figure 4.2: [13C-NMR] spectrum of Dinal

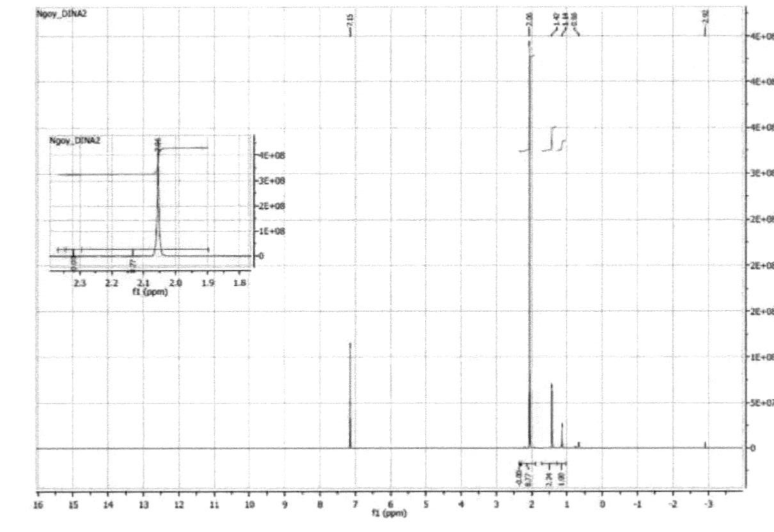

Figure 4.3: 1H-NMR spectrum of Dina2

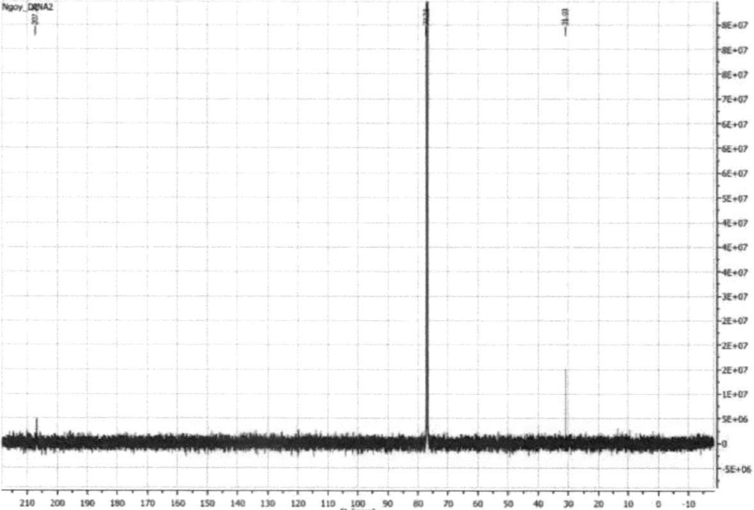

Figure 4.4: Dina2 13C-NMR-13C spectrum

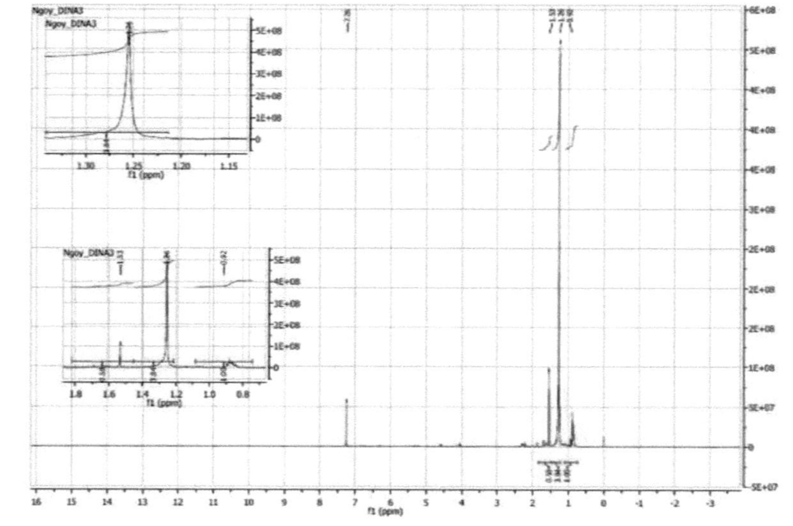

Figure 4.5. : 1H-NMR spectrum of Dina3

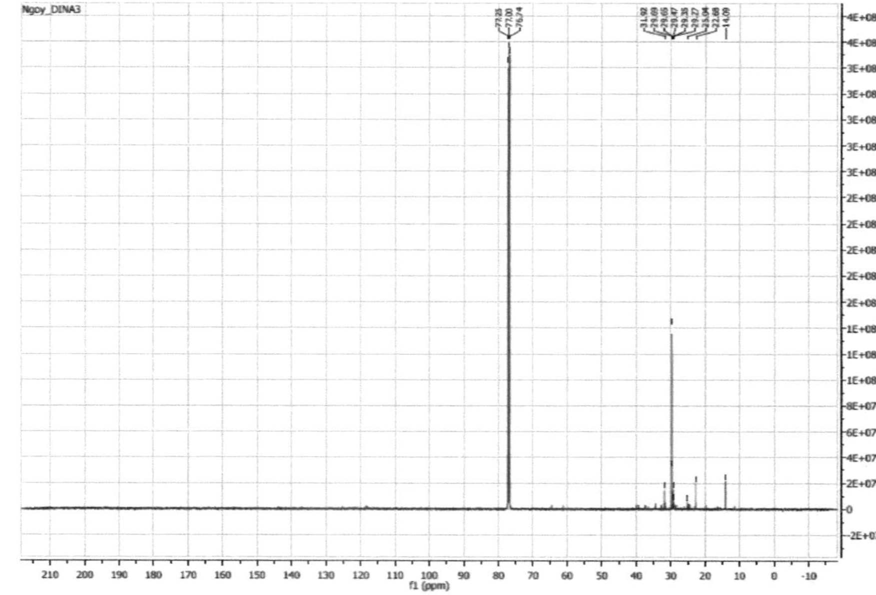

Figure 4.6: Dina3 13C-NMR spectrum

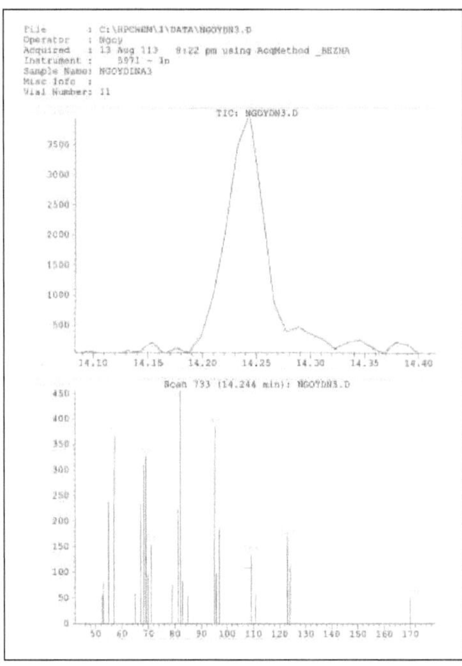

Figure 4.7: Chromatogram of the peak at 14.244min of Dina 3 and its Mass Spectra

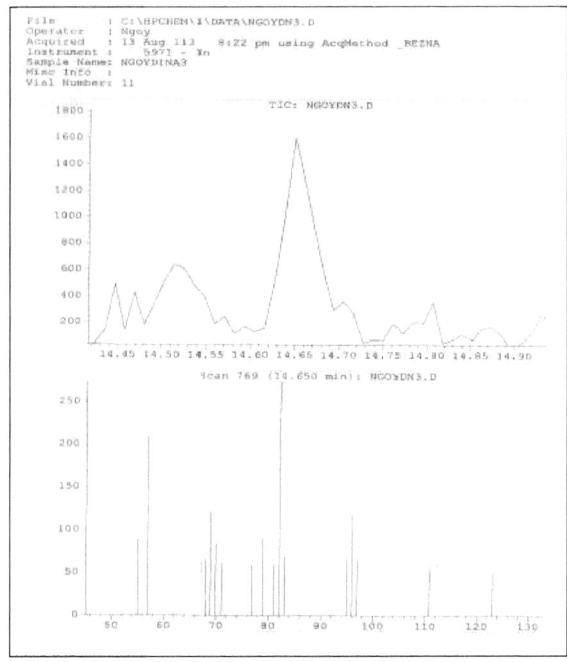

Figure 4.8. Chromatogram of the peak at 14,650 min of Dina 3 and its Mass Spectrum

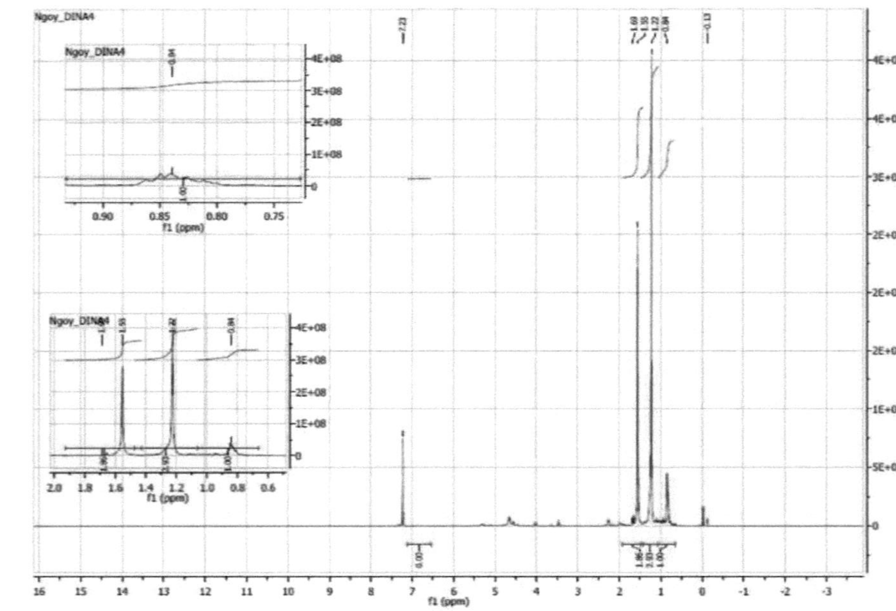

Figure 4.9: 1H-NMR spectrum of Dina4

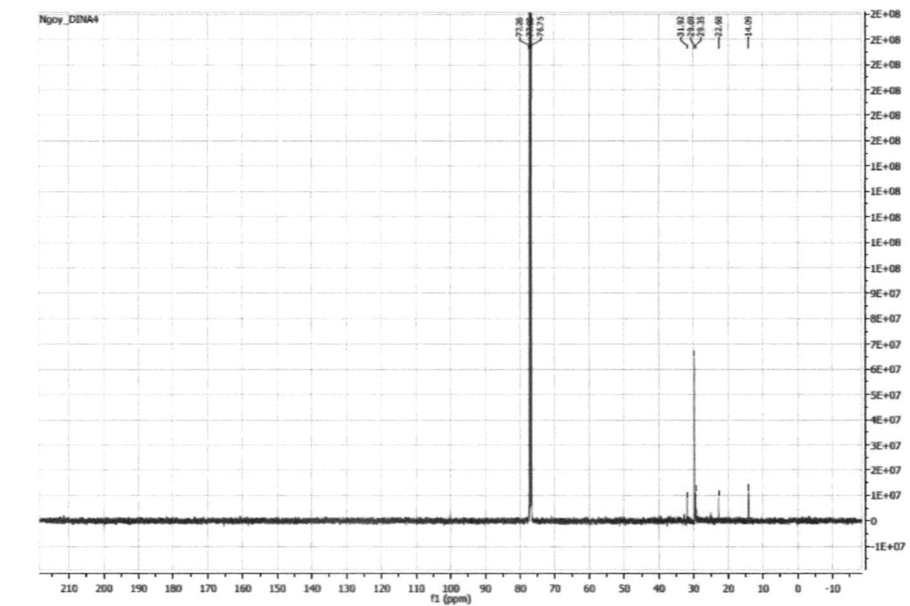

Figure 4.10.: Dina4 13C-NMR Spectrum

Library Searched : C:\DATABASE\WILEY275.L
Quality   : 83
ID   : 2-Phenyl-2-tipyl-acenapthenone

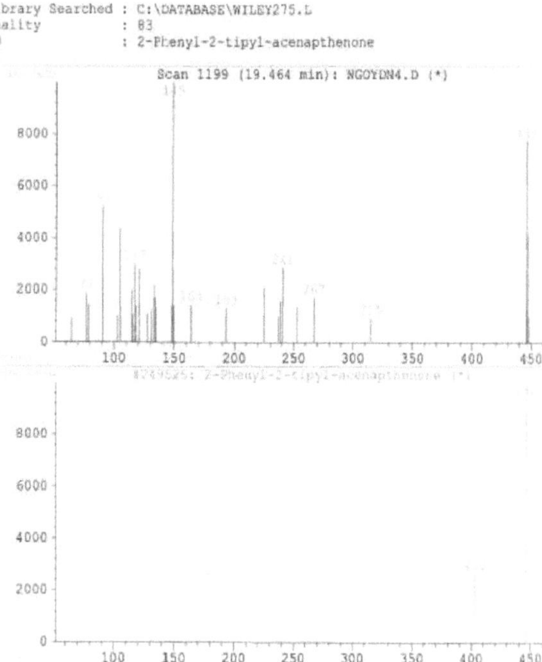

Figure 4.11.: Mass Spectrum of 2-Phenyl-2-tipyl-2-acenapthenone Dina4

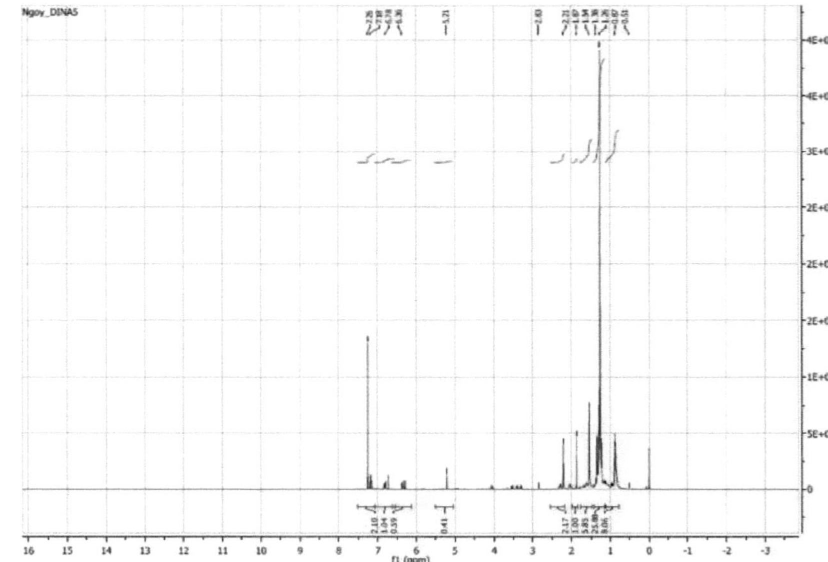

Figure 4.12.: 1H-NMR spectrum of Dina5

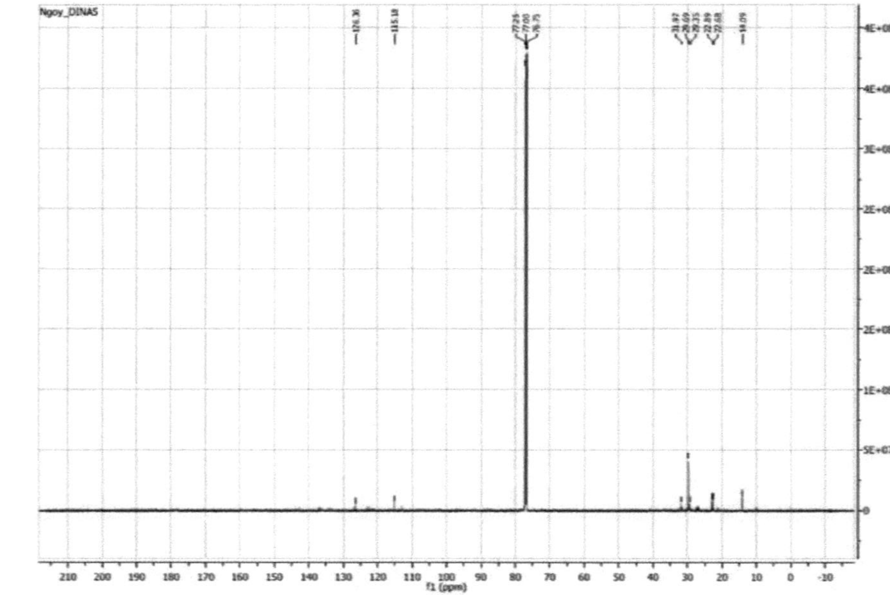

Figure 4.13.: Spectrum ^{13}C-NMR-13C of Dina5

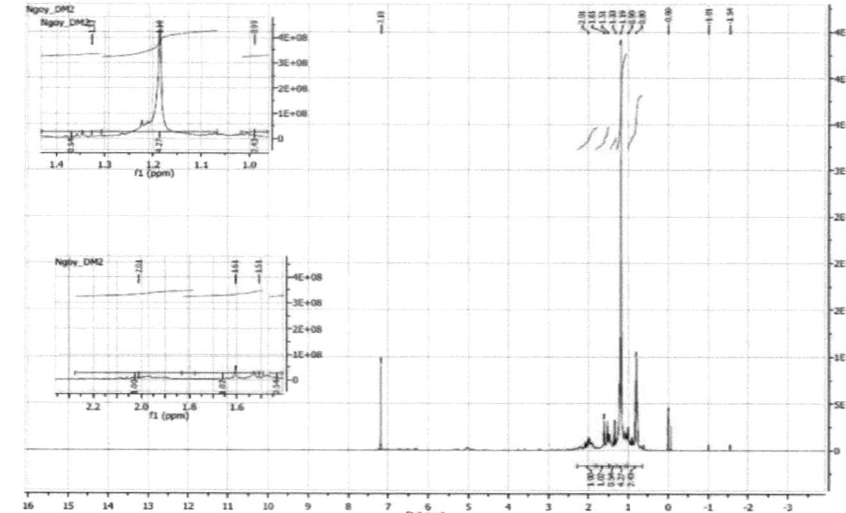

Figures 4.14.: [1H-NMR] spectrum of DM2

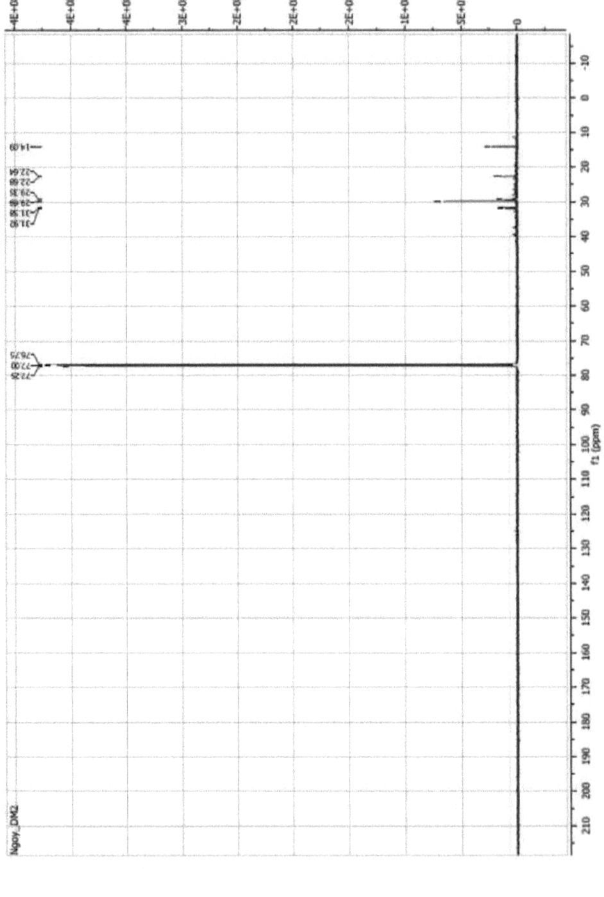

Figure 4.15.: ^{13}C-NMR spectrum of DM2

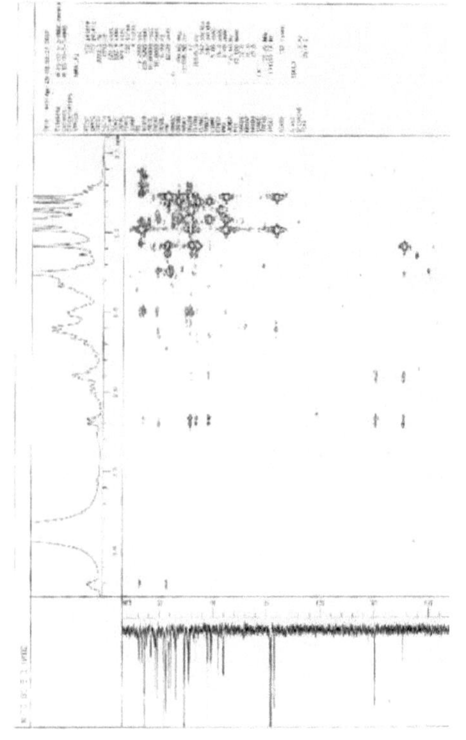

Figure 4.16.: HMBC-NMR of the majority Tsh-Do5/3/2 compound

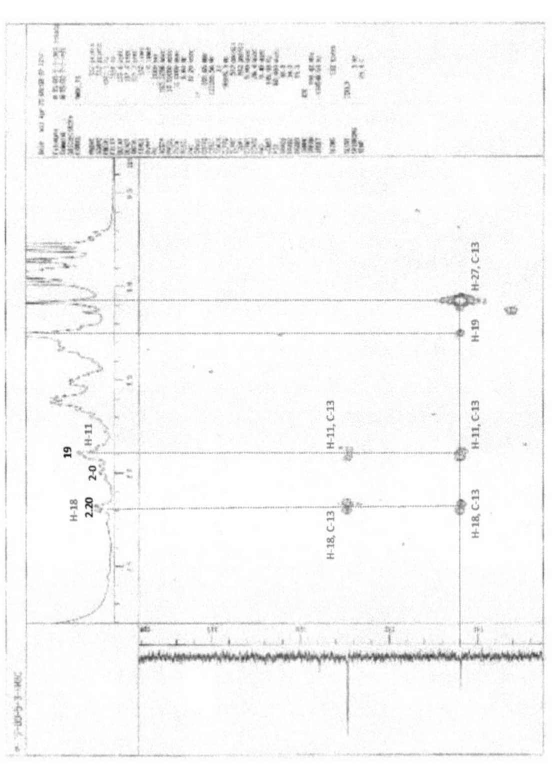

Figure 4.17.: HMBC-NMR of the majority expansion1 compound of Tsh-Do5/3/2

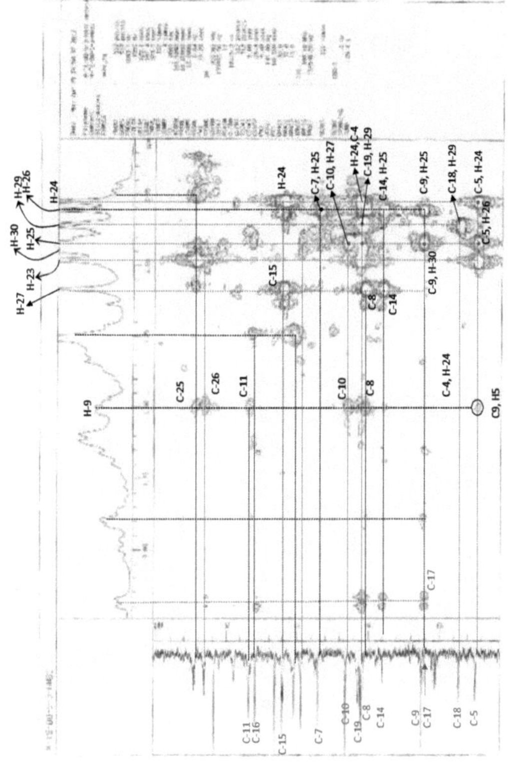

Figure 4.18. HMBC-NMR of the majority compound expansion 2 of Tsh-Do5/3/2

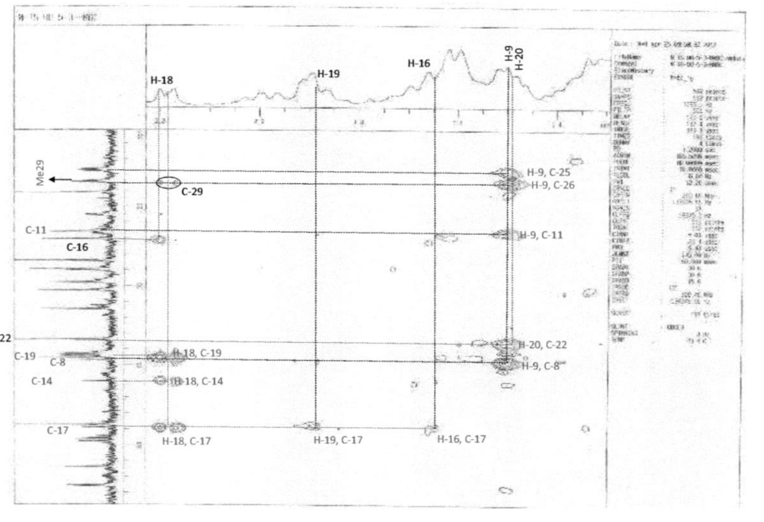

Figure 4.19. HMBC-NMR of the majority expansion3 compound of Tsh-Do5/3/2

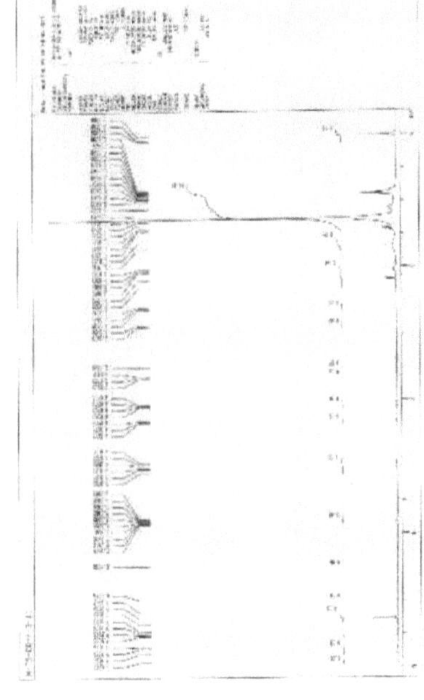

Figure 4.20.: ^{1}H-NMR spectrum of M-TS-DO-4-3

Figure 4.21. ^{13}C-NMR spectrum of M-TS-DO-4-3

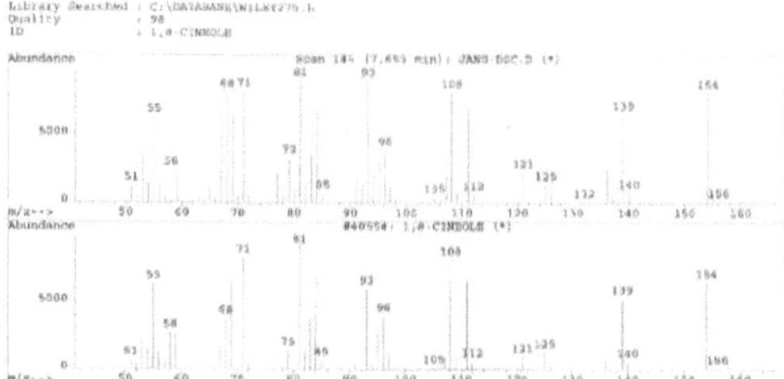

Figure 4.22. Comparison of the Do C and 1.8-cineole spectra

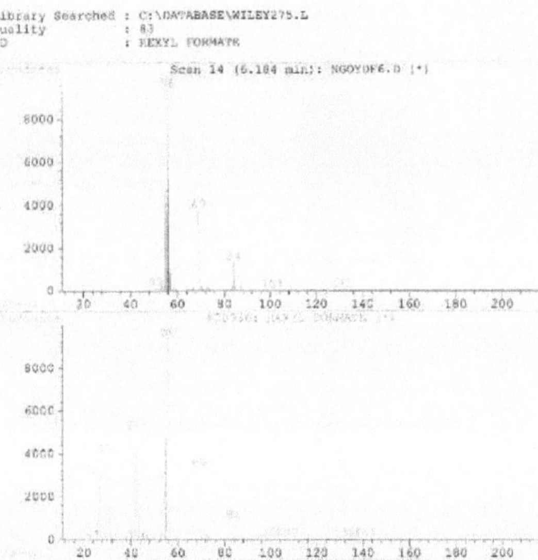

Figure 4.23.: Mass spectrum of hexyl formate and DF6

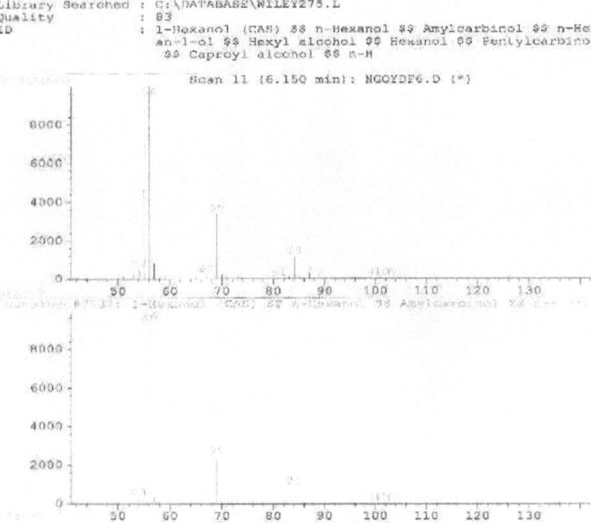

Figure 4.24. Mass spectrum of *n-hexanol* compared to DF6

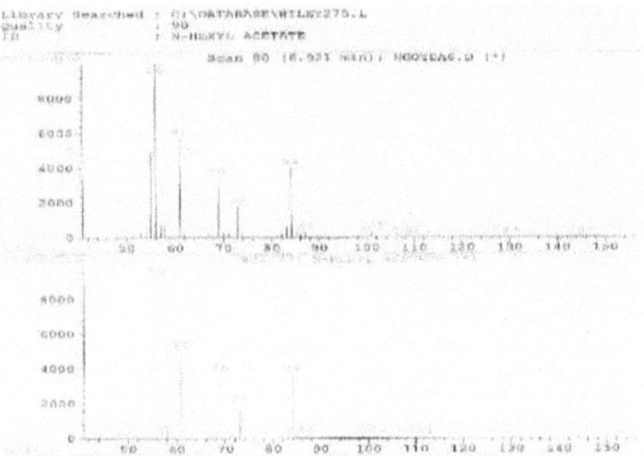

Figure 4.25. Hexyl acetate mass spectrum compared to that of DA6

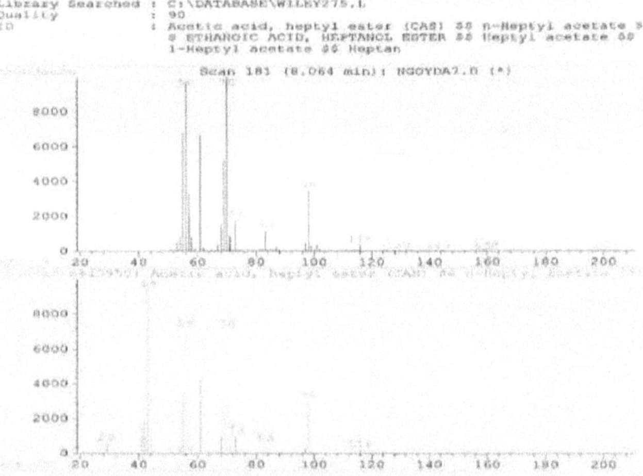

Figure 4.26.: Mass Spectrum of *n-heptyl* acetate compared to DA7

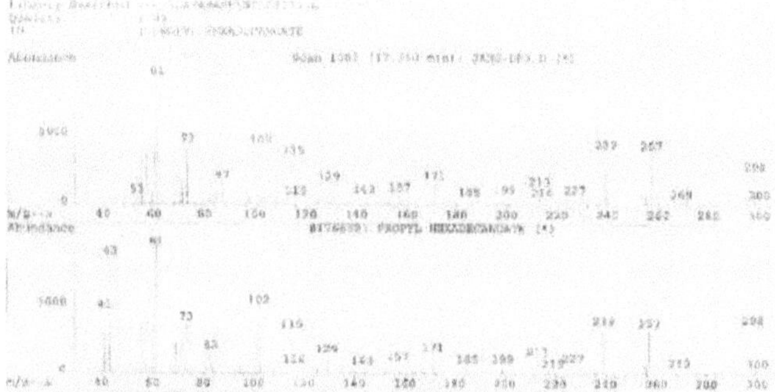

Figures 4.27. Mass spectra of *n-propyl* palmitate and DP3

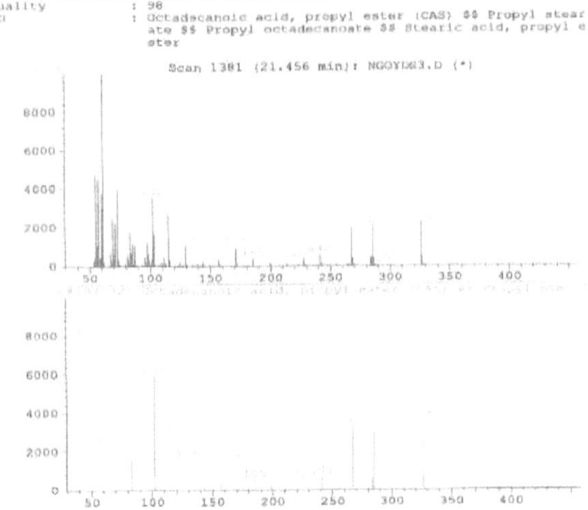

Figure 4.28. Mass spectra of *n-propyl* stearate and DS3

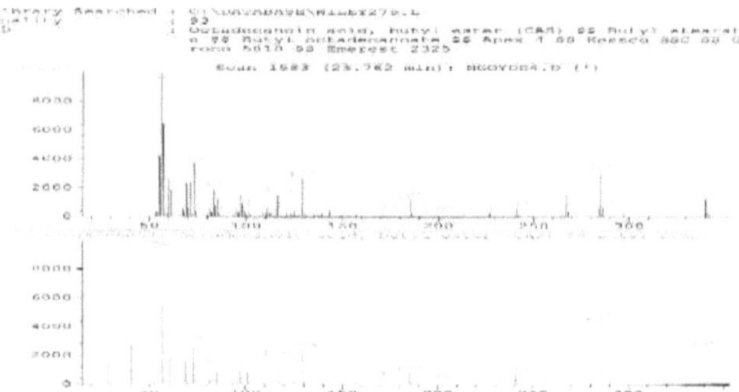

Figure 4.29. : Mass spectrum of butyl stearate compared to DS4

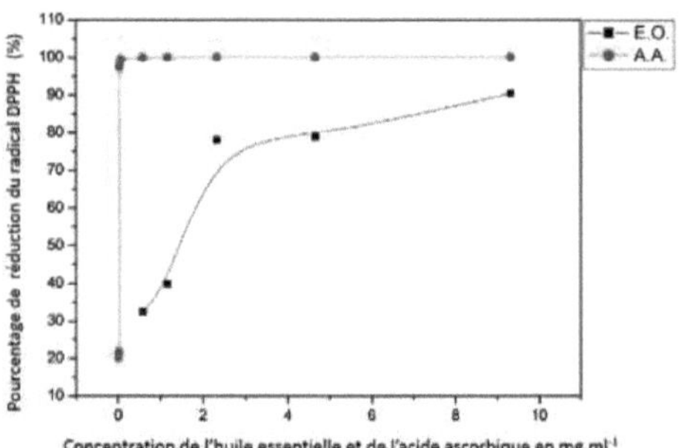

Figure 4.30. Comparison of DPPH radical reduction by *O. canum* essential oil (HE) and ascorbic acid (AA)

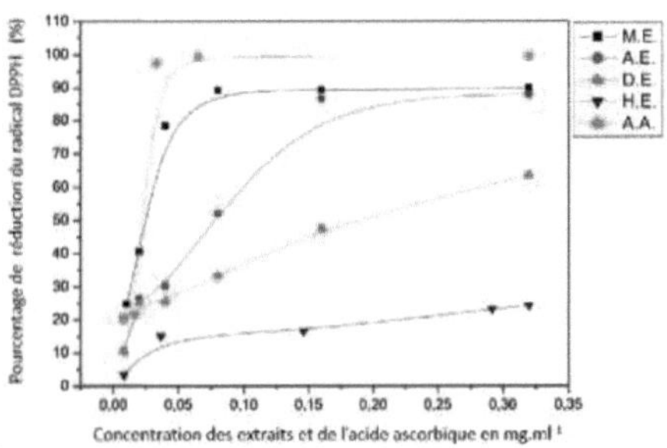

Figure 4.31. Variation in percentage reduction of the DPPH radical as a function of concentrations of *Ocimum canum* and AA extracts.

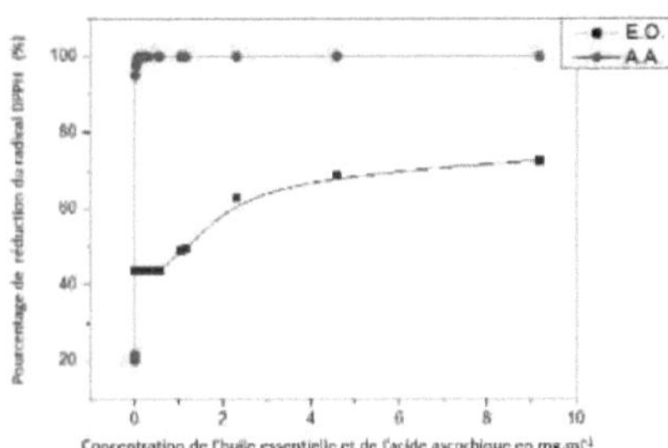

Figure 4.32. Comparison of the reduction of DPPH radicals by *O.basilicum* essential oil and ascorbic acid (AA).

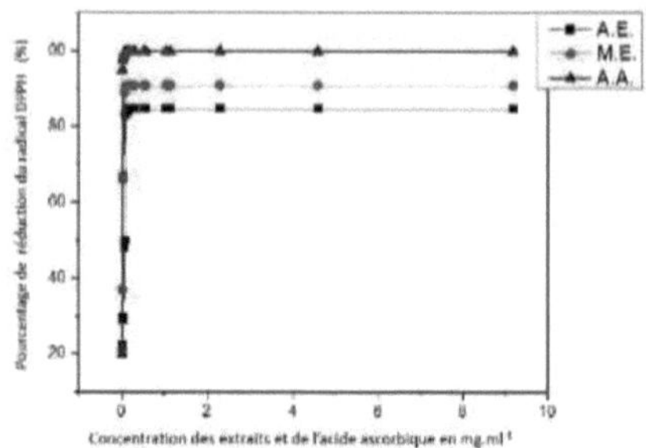

Figure 4.33.: Comparison of DPPH radical reduction by extracts (EM: methanolic extract, EA: ethyl acetate extract) of *O. basilicum* and ascorbic acid (AA)

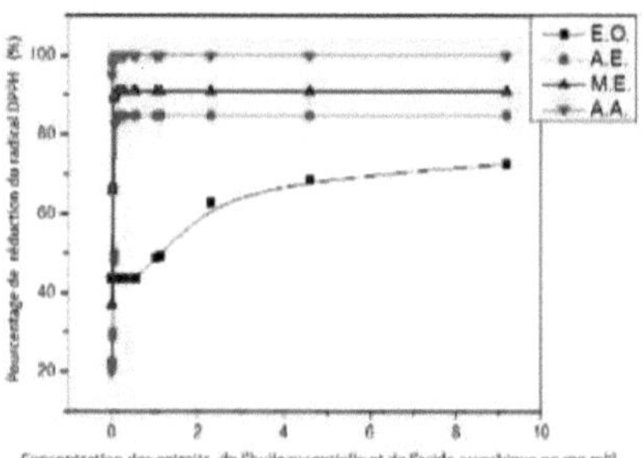

Figure 4.34: Comparison of the reduction of DPPH radicals by extracts (EM: methanolic extract, EA: ethyl acetate extract), by *Ocimum basilicum* essential oil and by ascorbic acid (AA).

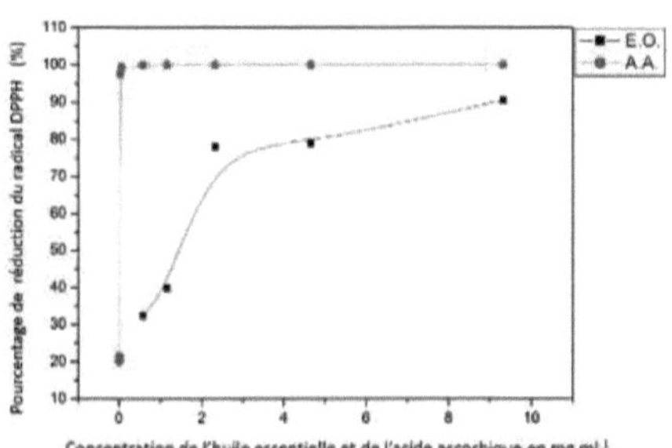

Figure 4.35. Reduction of the DPPH radical by the essential oil (HE) of *Ocimum gratissimum* and ascorbic acid (AA).

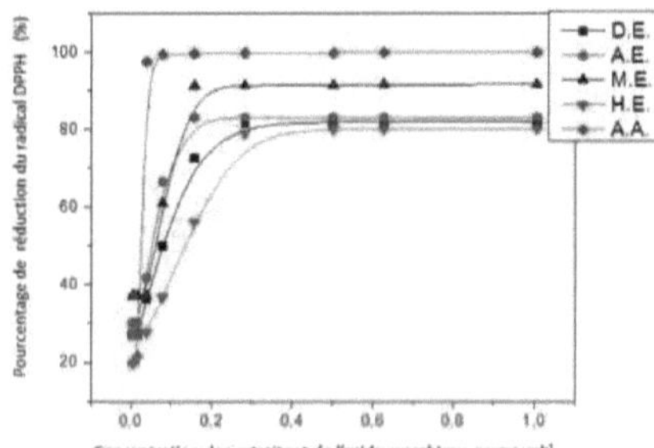

Figure 4.36.: Reduction of the DPPH radical by the four extracts of *O.gratissimum* (EM, EA, ED, EH) and by ascorbic acid (AA)

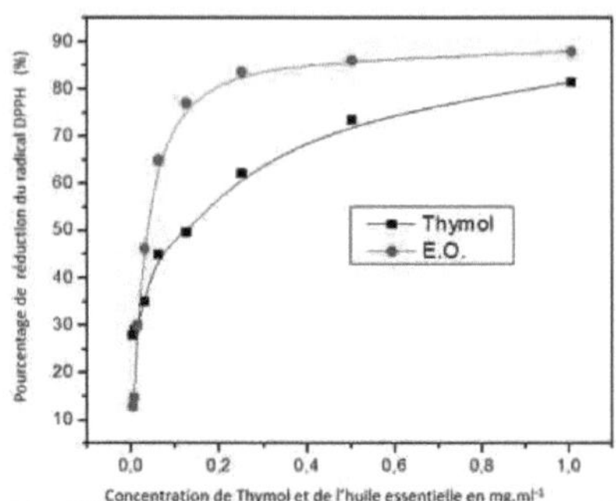

Figure 4.37.: Reduction of the DPPH radical by the essential oil of *O.gratissimum* (HE) and by pure thymol.

Table 4.1: Yield of fractional extraction of three plants

Plant	Mass of plant powder in grs	Extract	Mass extracted in grs	Return in % (%)
Ocimum basilicum	200,65	n-hexane	4,3562	2,171
		Dichloromethane	3,6065	1,7974
		Ethyl acetate	2,5489	1,270
		Methanol	3,766	1,877
Ocimum canum	100	n-hexane	0,589	0,589
		Dichloromethane	0,870	0,870
		Ethyl acetate	0,980	0,980
		Methanol	4,090	4,090
Ocimum gratissimum	1000	n-hexane	17,6296	1,763
		Dichloromethane	26,660	2.666
		Ethyl acetate	12,560	1,256
		Methanol	33,104	3,3104

Table 4.2: List of fatty acid esters found in the laboratory

Methyl esters of fatty acids found in the laboratory	
Methyl nonanoate	Methyl stearate
Methyl decanoate	Methyl nonadecanoate
Methyl undecanoate	Methyl docosanoate
Methyl tridecanoate	Methyl eicosanoate
Methyl myristate	Methyl heneicosanoate
Methyl pentadecanoate	Methyl tricosanoate
Methyl palmitate	Methyl tetracosanoate
Methyl heptadecanoate